GAS USAGE & VALUE

GAS USAGE & VALUE

THE TECHNOLOGY AND ECONOMICS OF NATURAL GAS USE IN THE PROCESS INDUSTRIES

DR. DUNCAN SEDDON

PennWell®

DISCLAIMER

The recommendations, advice, descriptions, and the methods in this book are presented solely for educational purposes. The author and publisher assume no liability whatsoever for any loss or damage that results from the use of any of the material in this book. Use of the material in this book is solely at the risk of the user.

1421 South Sheridan Road
Tulsa, Oklahoma 74112-6600 USA

800.752.9764
+1.918.831.9421
sales@pennwell.com
www.pennwellbooks.com
www.pennwell.com

Director: Mary McGee
Managing Editor: Marla Patterson
Production / Operations Manager: Traci Huntsman
Production Manager: Robin Remaley
Assistant Editor: Amethyst Hensley
Book Designer: Robin Remaley
Cover Designer: Charles Thomas

Library of Congress Cataloging-in-Publication Data Available on Request

Seddon, Duncan
Gas Usage & Value: The Technology and Economics of Natural Gas Use in the Process Industries

ISBN 1-59370-073-3

Printed in the United States of America
1 2 3 4 5 10 09 08 07 06

To Judith

Contents

Preface

The objective of this book is to give gas industry professionals an overview of the uses and value of gas to the process industries. The purpose of this book is to give an overview of the options and costs for utilizing natural gas. The first versions had the primary aim of helping companies to develop and value gas reserves by identifying possible uses for the gas and the kind of gas price a downstream industry will support. This book includes these features, which have been expanded and updated. It also includes work performed for a gas utilization workshop delivered by the author in Kuala Lumpur, which led to a gas marketing handbook that was written for BP-Amoco. This makes the book of wider appeal to anyone with an interest in developing, pricing, and selling gas.

1

Introduction

The current era has been called the *Age of Gas*. This replaced the previous era, the *Petroleum Age*, when oil produced all transport fuels and chemicals, which in turn had replaced the *Coal Age*.

The prospects for the current era look good. There are vast reserves of gas, and most countries in the world have some gas assets. Technologies exist that can utilize gas for all of the duties provided by petroleum fuels and coal, whether for the generation of thermal energy and electricity or for the production of chemicals. For some of these technologies, gas offers a significant technical and cost advantage over petroleum and coal.

However, the substitution of gas for petroleum or coal products in many fields has been slow or has only occurred in exceptional circumstances. For example, gas and LNG have been available since the 1960s, but still lag well behind coal as a primary source of electricity generation. By contrast, the conversion of the chemicals industry from coal usage to petroleum took less than two decades in the 1940s and 1950s.

The reasons for the relatively slow uptake of gas are several. Many of the major gas reserves in the world are remote from the major energy markets in the industrialized nations. The remote sites are often costly to develop, and also it is costly to bring the gas to market. Further, developments in petroleum processing have maintained a competitive advantage for liquid products, particularly in the transport fuel sector (e.g., the production of clean fuels). Also, coal remains the cheapest energy source and is available in vast quantities in many parts of the world, often closer to the required market in the industrialized nations. In addition, developments in clean coal combustion have largely maintained the cost advantage of coal versus gas in electricity production. Another point is that many of the gas-utilizing technologies are not as efficient as the older-era technologies. This results in negligible substitution or extensive government support (e.g., the use of gas for transport fuel).

In addition, natural gas is a fossil fuel, and the combustion emissions are subject to the Kyoto Protocol in some countries. Natural gas is experiencing some benefit relative to petroleum and coal from control of emissions. However, gas is now in competition with, and being displaced by, feedstock from sources other than fossil fuels, often with government support.

One of the reasons for the attraction of gas is the expected peaking of oil production over the coming decade. The timing and manner of the oil peak is still the subject of extensive debate within the petroleum community.[1] However, it is worth noting that some commentators also believe that gas production will peak in the not-too-distant future.[2]

This book offers an overview of the options and costs for utilizing natural gas. The earlier versions of the book had the primary aim of helping companies to develop and value gas reserves.[3] As mentioned in the preface, this book includes these features, which have been expanded and updated. It also includes work performed for a gas utilization workshop delivered by the author in Kuala Lumpur, which led to a gas marketing handbook, written for BP-Amoco.

The book critically compares the alternatives so that the most attractive options for gas use (and sales) can be identified for specific locations and conditions. For this purpose, technology capital and operating costs have been compared on the same basis (U.S. Gulf Coast location to a late 2004 cost base). From this, the production costs are estimated for various gas prices and compared to the traded prices of the products where appropriate.

The contents are as widely embracing as possible for viable technologies in 2005. From time to time, new technologies for using gas are identified, or more information on emerging technologies becomes available. These technologies include the conversion of gas into dimethyl ether (DME) or compressed natural gas (CNG) as alternatives to liquefied natural gas (LNG). It also includes improvements to the Fischer-Tropsch (FT) process (such as to produce lubricating oil base stock).

For brevity, the book does not cover technology still in the research and development stage. In particular, the extensive volume of material on the direct conversion of gas (methane) into ethylene or methanol has been omitted. Also omitted are those emerging technologies that require extensive government assistance in order to compete with conventional fuels. Of particular note here is the use of CNG in vehicle fleets.

Burning and Feedstock: Why Convert Gas?

Natural gas is used in two ways (fig. 1–1). It is either burned as a fuel to provide heat energy for some other use or it is used as a feedstock and converted into another salable product, such as methanol or ammonia. An example of an important use for burning is for the generation of electricity. This can then be sold into a utility distribution network or used for a dedicated end-user.

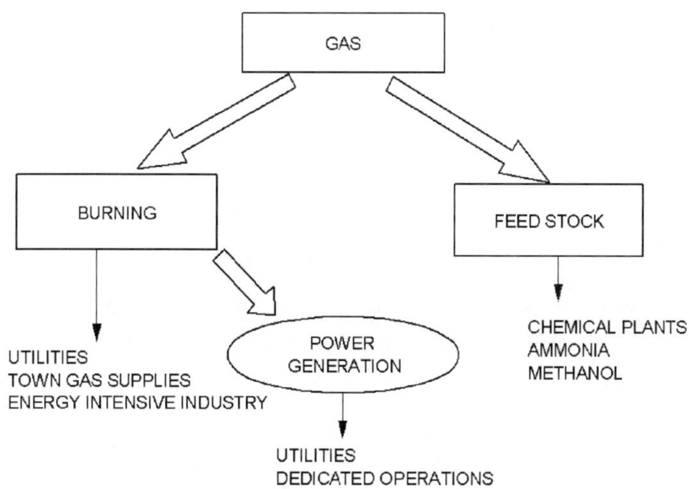

Fig. 1–1. Alternative uses for gas

For these uses, natural gas is in competition with other fossil fuels. For the most part, using gas is less capital intensive than using alternative fossil fuels, and the competitive position of gas is based on the trade-off between the usually higher price of gas and the lower capital cost required to use it.

However, in many parts of the world, massive reserves of gas exist with no apparent market for the fuel. For such reserves, there are two alternatives. The first is to use long intercontinental pipelines to bring the gas to the world population centers. The second alternative is to convert the gas into a transportable commodity, such as LNG or methanol, at or near the gas source.

In addition to these reserves, there are increasing quantities of gas available from oilfield developments. In the past such gas was flared, but increasing environmental concern is forcing oil exploration companies to consider every alternative for its use. Such associated gas can have a negative value. For instance, when flaring is prohibited, the oil cannot be recovered without the cost of reinjection. In other cases, flaring of the gas could result in payment of a tax. Associated gas is typically available in small quantities (10–50 MMcf/d). However, in some parts of the world (e.g., Nigeria and the Middle East), the available associated gas reserves are very large. Furthermore, in many cases the associated gas is offshore, which brings special problems if it is to be effectively utilized.

Units

A technical and economic appraisal of gas utilization opportunities spans several large subject areas: petroleum geology, petroleum and petrochemical refining, applied chemistry, chemical engineering, and economics. Unfortunately, these distinct fields carry their own units. Petroleum geology generally uses American units based on standards defined at 60°F. Most chemists and academic engineers use SI units. However, much petroleum engineering and refining technology uses a combination of both, often mixing American and SI units in the same function.

Also, as the fields of geology, petroleum engineering, and economics are crossed, emphasis switches from describing gas in terms of volume towards energy. Since the major use today of natural gas is in energy-intensive industries, this is understandable. However, carbon dioxide does not contribute towards energy, and gas of low energy content containing carbon dioxide may be an excellent feedstock for methanol production, where the carbon content may be more important.

This book employs units that the author considers to be the most widely acceptable, whether American or SI; where appropriate, an SI unit is given in parentheses. United States dollars of the day are used with a local equivalent in parenthesis as appropriate. Table 1–1 gives appropriate conversion figures. Note that there is a possibility of confusing U.S. 1,000 (M) with the SI mega (M). Units are spelled out where necessary to help avoid this confusion, and the designation *MM* is used to denote *million*.

Table 1–1. Units commonly used in the gas process industries

	U.S. Customary	S.I.
Volume	1 barrel (bbl	158.98 liters
	1 cubic foot (cf)	28.32 liters
	1 gallon	3.785 liters
Mass	1 pound (1 lb)	0.4536 kg
	1 long ton (2240 lb)	1.016 tonne
Energy	1 MMBTU	1.055 GJ
	1 MMBTU	293.1 kWh
Abbreviations		
1,000	M (mille)	k (kilo)
1,000,000	MM	M (mega)
1,000,000,000	MMM (billion)	G (giga)
1,000,000,000,000	trillion	T (tera)
1.00E+15	quadrillion (quad)	P (peta)
1.00E+18		E (eta)

Approximate Conversion Factors

Table 1–2 gives some approximate conversion factors that will help place the discussion of gas usage into perspective.

Table 1–2. Some approximate conversion factors

	MMcf/d	MMBTU/d GJ/d	TJ/d	PJ/a	tonne coal/d
Small unit processes	1	1,000	1	0.36	37
Local industry Small towns	10	10,000	10	3.6	370
Chemical plants Large towns and cities	100	100,000	100	36	3,700
World scale energy projects	1,000	1,000,000	1,000	360	37,000

The following approximate values are also useful. A list of energy values is given in appendix B.

- 1 tonne (t) coal = 27 GJ = 750 L diesel
- 1 tonne (t) diesel = 46 GJ
- 1,000 cf of natural gas is about 1MM BTU
- gas @ 1,000 BTU/cf = 37.5 MJ/m^3
- 1 MM BTU is approximately 1 GJ

Economic Analysis

The economic analysis follows the methodology described by the International Energy Agency (IEA) for evaluating alternative feedstocks for the production of petrochemicals.[4]

Gas utilization technologies are highly capital intensive, and some means has to be found for comparing alternatives. The method used for technology comparisons is to develop a fixed variable equation for a hypothetical greenfield plant producing a single product from natural gas feedstock. (A greenfield plant is a stand-alone operation comprising the main process units, often referred to as the battery limits, the utilities system, and the off-sites for product export. The only requirement is gas of suitable quality.)

The fixed costs of the plant are derived from published estimates of the costs of the capital items and operating costs. The operating costs can often be approximated as fixed percentages (typically 5% or 10%) of the total installed capital cost. The return on working capital is also included in the fixed-cost term. No account is taken of tax, depreciation, or allowances.

This leaves the natural gas as the only variable in the fixed-variable equation:

$$P = xNG + c \tag{1.1}$$

where

P is the product price,

NG is the gas price,

c is the constant representing the fixed costs, and

x is the quantity of gas

The largest component of the constant c is the return on investment of the fixed capital. In order to make comparisons easier, a standardized methodology is adopted, which is detailed in appendix F. Typically process plants are assumed to take a similar (three-year) construction period and to operate at full output over the project lifetime with zero residual value. The capital payback is then over this operating life.

Once the fixed variable relationships are derived, the equation can be used to estimate the production cost for any given gas price. By comparing the estimated production cost with traded prices for the product, the viability of a particular project can be determined. By considering alternative technologies at similar gas prices, alternative approaches to gasfield developments can be critically compared.

Approximations to the economic analysis

Fig. 1–2 illustrates the cost error for any given project as the project proceeds to completion. The error plus or minus is the error from the final cost, which is only known after the project is complete.

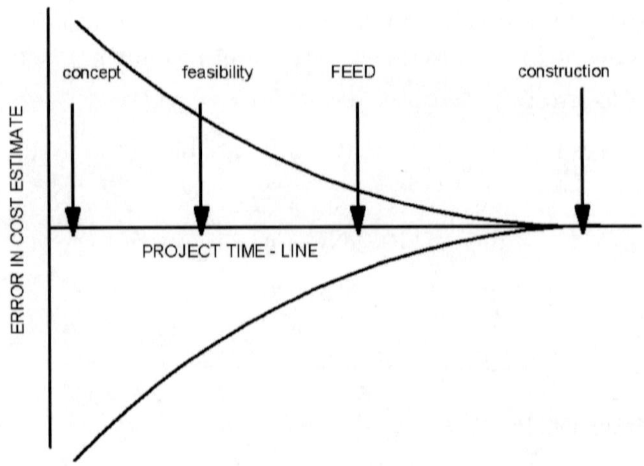

Fig. 1–2. Error in cost estimates

The first stage in the project is the concept study. This involves minimal expenditure in terms of the total project cost. The error in the cost estimate ranges from about ±25% to ±40% of the final project cost. The primary aim of this work is to improve the approach to the concept study so as to achieve an error in the lower end of this range. For many occasions (e.g., for very remote or unusual locations, such as offshore), this may not be feasible. The percentage of error may be as much as 100% or more.

The second phase is the feasibility study. This stage may require the expenditure of 1% to 2% of the total project cost. This will define the location, feedstock, and product market, as well as the technology to be used. It will also typically encompass an outline of regulatory approval and assessment of environmental impacts. The error in the estimate is typically not less than about ±10%. Financiers (bankers and corporate boards) like the error to be ±5%. This level of estimate can usually only be achieved by a front-end engineering and design (FEED) study.

A FEED study focuses on obtaining accurate cost estimates for the major items of process equipment and further definition of the most sensitive parts of the overall project cost. It is designated as a front-end study because it is performed prior to final corporate board and financial approval for the overall project. FEED costs are typically 5% to 10% of the overall project cost. Thus for a $300 million project, expenditure of at least $15 million will be required. Only after the feasibility and FEED studies will the cost error be in the vicinity of ±5%. In many instances, especially for new technology or for a remote location or offshore, the error will still be in the region of ±10% or more.

The full engineering design, procurement, and construction will also account for some error, hopefully < 5%. Only when the project is complete and running to the design specifications is the final cost of the project known.

As the adage goes, "the accuracy of the cost estimate is proportional to the time and money expended." Unfortunately, there is a common tendency to try to shorten or circumvent the costing process. This often leads to project failure.

Data Sources

Wherever possible, literature references are given. The *Oil & Gas Journal* articles are a useful source, and these often give further references to conference proceedings and articles published in the academic literature. As well as technical articles, *Hydrocarbon Processing* produces reviews of technology on a regular basis. Nowadays these are issued on a CD-ROM. These give more details of different technologies from the various process licensors.

In the descriptions of the various technologies, several assumptions and omissions to the process flow sheets have been made so as to help understanding of the principal issues and to improve the clarity of the description. If a particular technology or approach is of interest to a reader, then the process licensor should be approached for the latest updates and information.

References

1. See parallel articles by C. J. Campbell. "Industry Urged to Watch for Regular Production Peaks, Depletion Signals," and M. C. Lynch. "Petroleum Resources Pessimism Debunked in Hubbert Model and Hubbert Modelers' Assessment," *Oil & Gas Journal*, July 14, 2003, p. 38.

2. Imam, A., R. A. Startzman, and M. A. Barrufet. "Multicyclic Hubbert Model Shows Global Conventional Gas Output Peaking in 2019," *Oil & Gas Journal*, August 16, 2004, p. 20.

3. Seddon, D. "Gas Utilisation," Hindsford Pty. Ltd., 1998.

4. Statton, A., F. F. Hemming, and M. Teper. "Ethylene Production from Oil, Gas, and Coal Derived Feedstock," IEA, EAS Report E5, November 1983, and references therein.

2

Gas Availability

This chapter is principally concerned with opportunities for the establishment of world-scale plants that create products for an export market or a large domestic market. For instance, a modern methanol plant producing 2,500 t/d of methanol will require about 30 PJ/y (approximately 30 Bcf/y) of natural gas. With a lifetime of 20 years, it will require 600 PJ or about 0.6 Tcf. The following summarizes the current principal developments on a continent-by-continent basis. Obviously, some potential developments in specific countries are omitted in an attempt to provide an overview of the global situation.

The chapter examines the availability of natural gas on a regional basis. As well as natural gas, there is increasing interest in gas from alternative sources, such as coalbed methane (CBM) and natural gas hydrates (NGH). The status of these developments is also reviewed.

World Supply and Demand for Gas

The world natural gas reserves are more than 6,000 Tcf. This is very large considering the current annual consumption of about 100 Tcf. The world distribution of gas reserves at the beginning of 2005 is shown in figure 2–1.

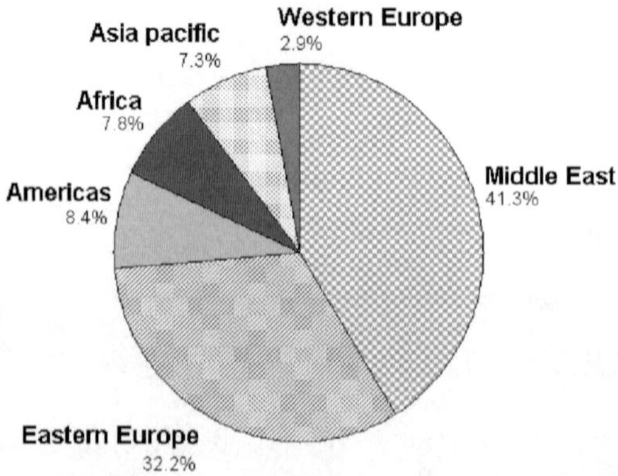

Fig. 2–1. World gas reserves (6102 Tcf). Source: M. Radler. "Worldwide Report," *Oil & Gas Journal*, December 20, 2004, p. 18.

As illustrated in the figure, most of the world supply is in Eastern Europe and the Middle East. The Eastern European reserves are dominated by Russian gas reserves (85% of the eastern European gas reserves), which exceed 1,680 Tcf. Some of the undeveloped fields may contain more than 1,000 Tcf of gas. In the Middle East, the major reserves are in Iran (940 Tcf). Qatar, in a series of gas fields known as the "North Field," has more than 910 Tcf of proven reserves.

In every part of the world, there is the potential to find large gas reserves capable of supporting a world-scale gas conversion plant. In areas near large urban populations, gas demand by these populations will tend to drive the gas price up. Sometimes this can result in the gas price being too high to support the use of gas as a feedstock for conversion. This is generally the case for gas discoveries in Europe and the United States.

Some areas of the world are regarded as being *gas-prone*, i.e., prospective oil basins will produce substantial quantities of gas rather than oil. This means that some parts of the world contain vast quantities of gas, far in excess of any domestic market demand. Such areas are important when considering the conversion of gas into another tradeable product such as LNG, methanol, or ammonia.

The world production and consumption of gas is shown in figure 2–2. Because gas is difficult to store, global annual production and consumption statistics are very similar. However, even though some regions are in balance, others have deficits, such as Western Europe and the Far East. These needs are filled by intercontinental transport of gas by pipeline or LNG from regions with excess gas, such as Eastern Europe and the Middle East.

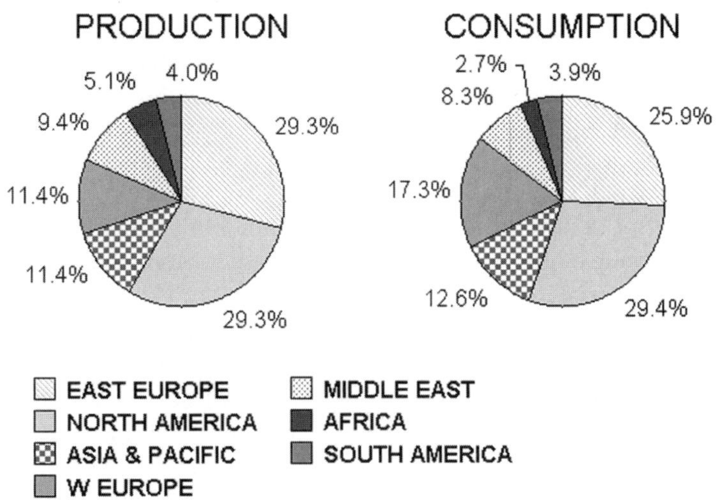

Fig. 2–2. World production and consumption of gas (2002 dry gas consumption: 92 Tcf). Source: IEA data.

Overview of Large Gas Reserves

Eastern Europe: Western and Central Russia

The Eastern European gas reserves for 2005 are shown in figure 2–3.

In any consideration of the use and conversion of natural gas, the enormous reserves in Russia must be considered. These are estimated at 1,680 Tcf (35% of the world's proven gas reserve). Most of the present annual production (about 20 Tcf) comes from four deposits: Urengoi, Yamburg, Medvyezhe, and Vyngapur in West Siberia. Vast amounts of this gas are moved by intercontinental pipelines to fuel the industries of Eastern and Western Europe.

The mass transport costs of this gas and its conversion will profoundly influence the value of any commodity chemicals such as ammonia. In the days of the Soviet Union, the enormous size of the gas reserves, coupled with the integrated production and shipping abilities of a centrally planned economy, resulted in Russian domination of the world ammonia market. With the breakup of the Soviet Union, and the separation of production and shipping facilities between different countries, this domination has lessened somewhat. Nevertheless, any gas conversion program will have to take into account the potential of Russia's entry into the market and its influence on the traded price of the intended product.

The outlook for Western Russia as a major world supplier of gas and gas derivatives is good. This outlook is aided by the integration with gas developments in the Caspian Sea region and the very large undeveloped super giant fields in the Barents Sea.[1]

Although the Russian gas reserves dominate the Eastern European sector, some of the other independent states have significant undeveloped reserves. Reserves for Kazakhstan are 65 Tcf; Turkmenistan, 71 Tcf; Ukraine, 39 Tcf; and Uzbekistan, 66 Tcf. Much of this gas is located around the Caspian Sea.

Fig. 2–3. Eastern European gas reserves (1,966 Tcf)

Middle East

The gas reserves at the beginning of 2005 in the Middle East are illustrated in figure 2–4.

The Middle East contains enormous gas reserves (2,522 Tcf). The development of the super giant gas fields of Qatar (910 Tcf) and the United Arab Emirates (Dubai, 196 Tcf) are important to the immediate world gas markets. Future developments would be in Yemen (19.9 Tcf) and Iran. In size, Iran's gas reserves (940 Tcf) rival those of Russia but as yet remain largely undeveloped. In progress is the development of the large South Pars field that is slated to supply gas to new chemicals plants.

Saudi Arabia has more than 234 Tcf of gas. Until recently, most of the gas production was from reserves associated with the nation's very large oil production. However, recently two very large gas plants (1.5 Bcf/d) have come onstream that use nonassociated gas.[2] Saudi Arabia has embarked on the extensive use of its gas reserves to produce chemicals.[3] Sales gas is used to supply major industrial sites and to produce downstream products such as ammonia and methanol.

The value of gas is generally very low (almost zero). Some countries in the Middle East have targeted gas-intensive conversion industries as a means of developing and selling gas.

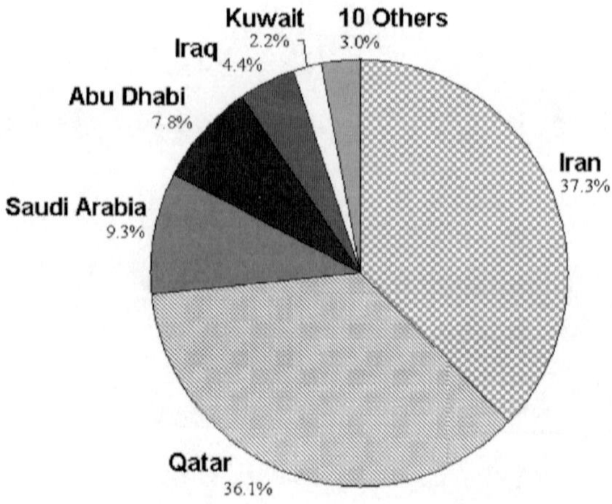

Fig. 2–4. Middle East gas reserves (2,522 Tcf)

Africa

The distribution of African gas reserves in 2005 is illustrated in figure 2–5. Africa has a total of nearly 500 Tcf of gas, which is concentrated in Nigeria, Algeria, Libya, and Egypt.

There are extensive gas reserves in North Africa, particularly in Algeria (160 Tcf), Egypt (58 Tcf), and Libya (52 Tcf). This supports extensive LNG trade in Algeria and pipeline transmission to the European Union. Gas in Egypt has been rapidly developed over the past decade to support the local economy.[4]

Apart from Nigeria (176 Tcf), where very large gas reserves are associated with its large oil fields, central Africa is largely unexplored. To date, much of the Nigerian associated gas has been flared. This waste has led to the establishment of an LNG plant and proposals for a large gas-to-liquids (GTL) facility.

One area of increasing interest is offshore Angola, where the search for oil has revealed several large gas resources. This has brought proposals for an LNG plant.[5]

In South Africa, the conversion of gas into synthetic fuels at Mossel Bay was planned and executed in the Apartheid era.

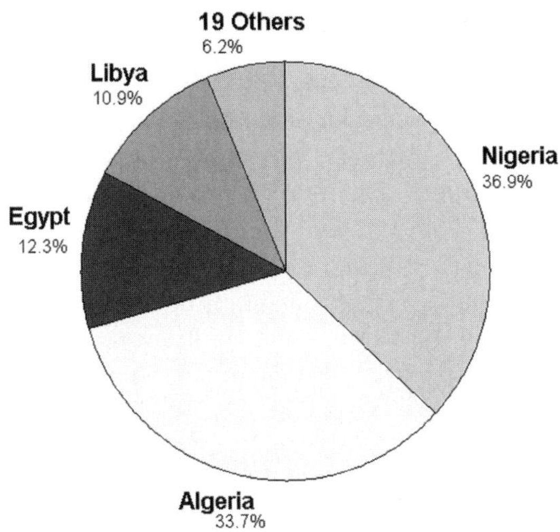

Fig. 2–5. African gas reserves (476 Tcf)

North and South America

The 2005 gas reserves of North and South America are illustrated in figure 2–6. The majority of these reserves are located in Canada, the United States, and Venezuela.

South America. There is extensive interest in developing the gas reserves in order to spur industrial development, particularly power generation. In the south part of the continent, Argentina (22 Tcf), Bolivia (24 Tcf), and Brazil (9 Tcf) are connected by pipeline.[6] In the north, Venezuela (151 Tcf) has increasing gas developments and an expanding gas pipeline system.[7] Trinidad (25 Tcf) has extensive chemicals operations using gas as feedstock, which includes LNG, methanol, and ammonia. Several other world-scale projects are under consideration.[8] Gas in the more remote regions is being piped to the population centers, and some gas, as in southern Chile, is used to produce methanol. In some areas (e.g., Peru, 9 Tcf), large gas reserves have been identified but as yet remain undeveloped. Here proposals include LNG and NGL-GTL schemes.[9]

North America. The North American gas market is dominated by demand from the population centers of the United States (annual consumption exceeds 22 Tcf). There is extensive gas trade from major gas fields in central Canada and Mexico to the United States.

In the north, the central Canadian fields of Alberta and Saskatchewan are very large. As well as processing for pipeline gas, much of it is destined for U.S. markets. There is also processing to produce fertilizer and methanol.[10] This market has all the attributes of a truly competitive market. There are many players in gas supply, transmission, and intermediate gas storage. Government regulation is aimed at maximizing competition, rather than granting monopoly rights, as is the case in many parts of the world.

To the south, the Mexican pipeline system is set to continue to expand to bring more gas into the southern states of the United States. In addition, the gas will be used to feed power plants throughout Mexico.[11] In the far north of the continent, large gas fields in Alaska remain largely undeveloped. Some development to produce LNG and fertilizer has occurred in gas fields in the Cook inlet, and development of the remoter fields could occur if the two areas became interconnected by pipelines.[12]

As well as producing LNG in Alaska, the U.S. imports LNG through four terminals in the Eastern states (annual capacity approaching 1 Tcf, with imports of typically about 200 Bcf). The import demand is primarily to supplement domestic and pipeline supplies in periods of high demand (and price).[13] There are plans to increase the import of LNG by establishing terminals on the West Coast.

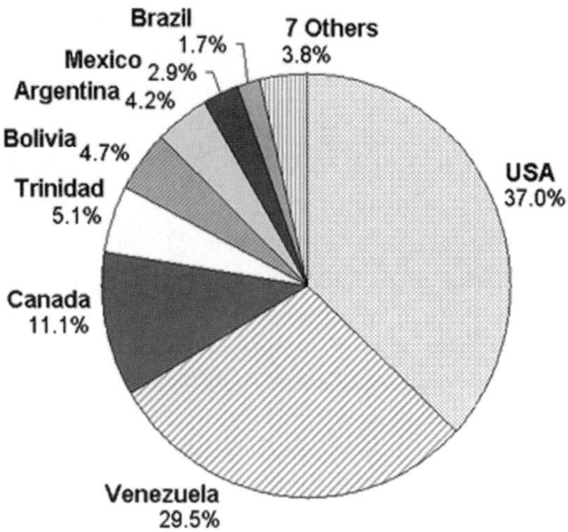

Fig. 2–6. Gas reserves of North and South America (511 Tcf)

Asia Pacific

The breakdown of reserves in the Asia Pacific is shown in figure 2–7. This illustrates that the region has more than 445 Tcf of gas. The principal reserves are in Indonesia (92 Tcf), Australia (90 Tcf), and Malaysia (75 Tcf).

Indian subcontinent. The past decade has seen major growth in gas discoveries and developments in India (30 Tcf), Pakistan (27 Tcf), and Bangladesh (10 Tcf). The Oil & Natural Gas Corporation and the Gas Authority of India Ltd. (GAIL) have been instrumental in the development of several gas fields in India. The production has found ready markets, which has resulted in a pipeline distribution system from Gujarat to Uttar Pradesh in the northwest region of India.

However, India is considered deficient in gas reserves, and there is a rapid growth in the demand for both gas and power. Existing fields produce about 3 Bcf/d, which just meets demand in 2005.[14] However, demand growth is rapid, estimated to be more than 5 Bcf/d by 2012. There are active developments to use LNG at four coastal sites around India and to deliver gas and power to the expanding economy.[15]

Recent discoveries of natural gas in Pakistan have eliminated the need to consider imports from Iran. Natural gas is replacing fuel oil as the method for power generation. It is also used as a feedstock for fertilizer (ammonia) production and is distributed for residential use.[16]

The gas industry in Bangladesh has grown rapidly over the past decade and now is a major contributor to the economy. By comparison to India, Bangladesh has an excess of gas over domestic demand, and there are proposals to supply the eastern coast of India and eventually link into the northwest gas system.[17]

Southeast Asia. All countries in Southeast Asia are increasing their use of gas. Initially gas developments are focused on supplying local domestic markets (Malaysia and Thailand), which can command high prices to the gas suppliers. Some counties (Australia, Brunei, Indonesia, and Malaysia) have large reserves that are remote from population centers. This has focused attention on developing the large Far East LNG trade, which supplies gas to Korea and Japan (countries that lack significant gas reserves).

Indonesia (90 Tcf) has very large gas fields that are used to feed LNG operations. The first large gas plants were built in the Aceh region of northern Sumatra. These fields have now peaked, but major expansions in LNG capacity have occurred in East Kalimantan on Borneo. Two major fields of potential interest are the Natuna and West Natuna fields in the South China Sea. The Natuna field is very large (said to contain 200 Tcf) but suffers from a very high carbon dioxide (CO_2) content (said to be approximately 70%). This results in difficult processing. The West Natuna field has been developed to supply Singapore with gas via a 650-km undersea pipeline. This pipeline network could be part of a mooted Southeast Asian gas grid that would interconnect fields in Malaysia, Thailand, and Indonesia, and possibly Australia. Yet another potential Indonesian development is a major field in the West Irian province. This has the potential to form the site of another LNG exporting facility.[18]

New Zealand (1.1 Tcf) has established large gas conversion plants producing products used for domestic consumption (fertilizer) or export (methanol). There has been discussion recently that the present reserves are insufficient to maintain the current operations.

Papua New Guinea (12 Tcf) has large gas fields as yet to be developed. One major project is the supply of gas to Australia via an undersea pipeline and a long (3,000-km) pipeline to the markets in southern Queensland.

There is extensive and growing use of gas in Thailand. The main producing fields are in the Gulf of Thailand. The production from these fields is transported via two pipelines to the Bangkok region to power the nation's rapidly growing industries. This gas supplies about 70% of Thailand's demand, the remainder being satisfied through imports from Burma. There are proposals to link with the Malaysian pipeline system and to supply both Malaysia and Thailand from fields under development in the Gulf. Most gas (more than 60%) would be used for power generation in place of oil- and coal-fired systems. Gas demand in 2005 is estimated at about 3 Bcf/d and is expected to rise to more than 5 Bcf/d by 2015.[19]

Burma (Myanmar, 10 Tcf) has developed a large gas field in the Andaman Sea (Yadana). This supplies gas to a power plant in Thailand but would remain separated from the main gas developments in the Gulf of Thailand that provide gas to Bangkok. This development has survived strong local opposition and contractual disputes.[20] Plans to use the gas for industrial developments in Burma have so far not materialized.[21]

In the Philippines, a gas project has been established that pipes gas from fields in the South China Sea to power generation plants on Luzon.[22]

Vietnam (7 Tcf) is now utilizing the associated gas of the Bach Ho oil field.[23]

Northeast Asia. Taiwan, Japan, and Korea have only small indigenous gas reserves. Their gas-based industries are driven by very large LNG imports.

China has very large gas reserves (53 Tcf), which are largely undeveloped. In 2002, of 660 cities, 649 had access to liquefied petroleum gas (LPG), but 163 cities used coke oven gas, with only 120 using natural gas. However, due to the greater environmental benefits of gas and the increasing demand from the expanding economy, this is set to radically change.[24] The Chinese annual demand for gas is estimated at 64.5 billion cubic meters (bcm) (2.3 Tcf) in 2005, and is expected to increase to 252 bcm (8.9 Tcf) by 2020.[25] This expansion will be achieved by both the development and transmission of the western gas reserves and by LNG imports on the eastern seaboard.

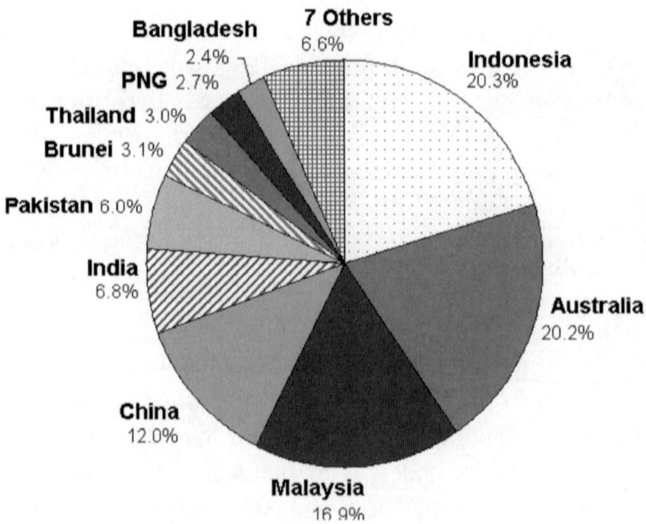

Fig. 2–7. Asia-Pacific gas reserves (445 Tcf). Source: "Worldwide Report," *Oil & Gas Journal*, December 2004. The data have been corrected for Australia.

Western Europe

The 2005 gas reserves of Western Europe are illustrated in figure 2–8.

The economies of the European Union (EU) utilize substantial volumes of gas for industry and power generation. The increasing use of gas has been at the expense of indigenous coal production. In the EU, only the United Kingdom (21 Tcf) and the Netherlands (62 Tcf) are self-sufficient in gas production (from associated gas and gas fields of the North Sea). Large volumes of gas are imported into the EU from Norway, North Africa (Algeria and Libya), and in particular Russia, which also supplies much of Eastern Europe with gas.

Norway has very large offshore reserves (74 Tcf) that have been developed to produce methanol as well as export to the EU.[26]

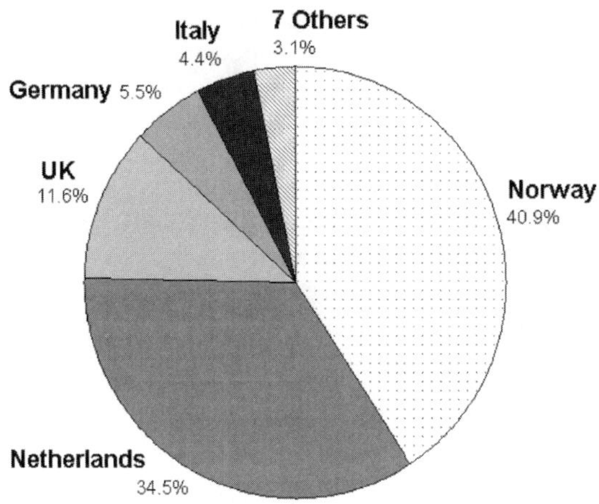

Fig. 2–8. Gas reserves of Western Europe (180 Tcf)

Alternative Sources of Gas

Gas from in situ coal: coalbed methane (CBM)

Recently there has been an upsurge in obtaining gas (methane) directly from coal seams. The gas found in coal depends on several factors, including coal rank, coal-in-seam pressure (usually associated with an aquifer), and local geological differences. As a guide, a 200-m deep, low bituminous coal may contain 10 m^3/t of extractable gas.[27] The vast coal reserves of the world make the potential of coal seam gas appear enormous. Further, coal is more widely distributed than gas, which implies that gas from coal would be available in those regions deficient in gas.

As an example, the CBM resources of Indonesia have been estimated at 337 Tcf, compared with natural gas reserves of 90 Tcf.[28]

The extraction technique is to drill into the seam from the surface and remove water, which is usually present and is holding the gas in place. This may take several months of mechanical pumping. The well can then be used to produce the gas. In order to maintain a constant gas supply, producing fields comprise many wells, some dewatering, some producing, and some in decline. However, environmental impacts have been identified as a cause for concern, particularly with regard to pollution from the extracted water.[29]

This production method is generally of higher cost than that for producing gas in the conventional manner, and as a consequence, coal gas is generally a high-cost gas. Tax incentives can lower this cost. The most extensive use of CBM is in the United States, where tax incentives are available.

In some underground collieries, methane drainage has to be undertaken for safety reasons. Since such gas is a necessary by-product of coal production, it can be available at a low price.

Natural gas hydrates

Natural gas forms a hydrate with water at suitably high pressures and low temperatures.[30] Vast reserves of hydrates have been identified in cold, deepwater environments; reserves in the range of 300 Tcf are often found. The hydrate layers often lie 300 m or so below the seabed, trapped between an impervious layer and a water layer. Table 2–1 lists some of the identified offshore hydrate fields and their salient properties.

Hydrates are metastable compounds that are stable at high pressures and low temperatures. Figure 2–9 illustrates the position of hydrate equilibrium of interest. This illustrates the boundary between free gas (lower right-hand side) and the areas where hydrate is stable. As can be seen, gas hydrates are stable over a wide range of temperatures and pressures.

Table 2–1. Some natural gas hydrate resources (after Makogon, Holditch and Makogan 2005)

Location	Water Depth (m)	GHL (m)	PD (bar)	Hydrate T (C)
Black Sea	2020	2030–2040	160	4.0
Blake Ridge–1	2790	2990–3220	200	11.0
Blake Ridge–2	3500	3600–3700	20	22
Bush Hill	2420	2440–2480	95	4.0
Costa Rica	3100	3400–3439	260	10.0
Guatemala–1	2400	2750–2800	125	15.6
Guatemala–2	2000	2450–2500	27	18
Guatemala–3	1720	1870–2120	125	9.5
Japan Sea	2600	2600–2650	95	17.0
Mexico–1	1950	2050–2212	125	7.0
Mexico–2	2900	3000–3077	250	5.2
Mexico–3	1950	2050–2212	130	7.2
Mississippi Canyon	1330	1365–1470	115	7.0
Nankai–1	945	1141–1210	45	11.0
Nankai–2	4700	4800–4870	415	4.0
Peru–Chile–1	5070	5200–5260	430	6.5
Peru–Chile–2	3900	3950–4000	305	10.0

GHL = gas hydrate layers; PD = pressure drop required to dissociate the hydrate

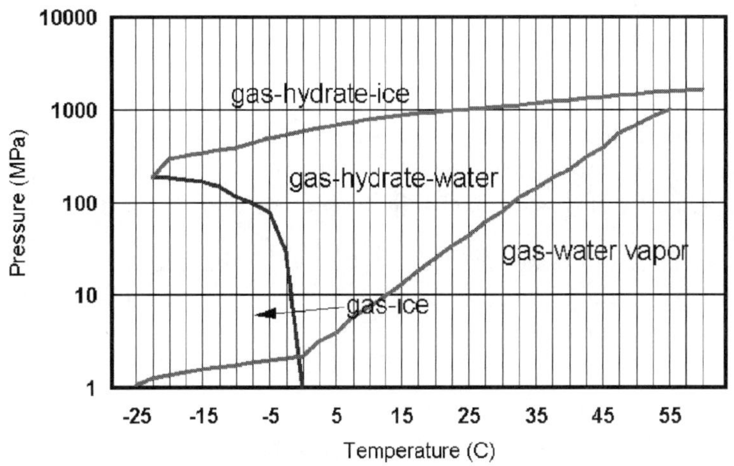

Fig. 2–9. Methane hydrate equilibrium. Adapted from Makogon, Holditch, and Makogan. "Russian Field Illustrates Gas-Hydrate Production," p. 43.

Given a hydrate source, in order to develop the reserve, the temperature and pressure of the reserve have to be adjusted to a position outside the equilibrium line (lower right-hand section of the figure). There are three alternative approaches to developing the reserves:

- **Reduce the hydrate pressure below the hydrate equilibrium pressure.** For many reserves, this requires a very large pressure drop. This appears only suitable for a minority of reserves where the equilibrium pressure is relatively low.
- **Increase the reservoir temperature above the hydrate equilibrium temperature.** This is the most applicable. However, the methodology of achieving this is still the subject of laboratory research and field trials.[31]
- **Inject chemicals to break the hydrate and separate the methane from water.** This method is used for the reverse process, namely the prevention of hydrate formation in cold and water-wet environments including gas pipelines. However, for gas hydrate field development, the method appears prohibitively expensive.

There are major development programs underway that may, in future years, result in gas being developed from these reserves.[32]

Tight gas

The term *tight gas* is used to cover gas in low-permeability sandstones. These reserves are thought to be very extensive, with many hundreds of trillion cubic feet of gas potentially available in the United States, mainly around the Rocky Mountain region. However, the extent and availability of this reserve have been questioned.[33]

Because production rates are restricted by the permeability of the sandstone, production from such zones is costly and often attracts government subsidies.

Synthetic natural gas (SNG)

The technology for the production of synthetic natural gas (methane) from other fossil fuels dates from the early 1930s. The development parallels the development of the Fischer-Tropsch process, central to GTL technology.

The process follows the same basic route and is detailed more fully in later chapters. The fossil fuel is first gasified to synthesis gas (carbon monoxide and hydrogen). This synthesis gas is then cleaned, which is an extensive operation for many of the raw gases produced by coal gasification. This can produce a wide range of by-products. A process known as water gas shift then converts some of the carbon monoxide to hydrogen. The gas is then passed over a nickel-based catalyst to produce methane by the methanation reaction. The background literature is very extensive and is not further developed here.

The production of SNG was widely practiced in Europe and in other countries during the 1950s. It was largely displaced by the widespread introduction of natural gas during the 1960s and 1970s.

However, a large lignite plant was built in North Dakota in 1984 at a cost of $2.1 billion. This plant produces SNG that feeds into a main gas pipeline delivering gas from Canada to the northern states of the United States. Some statistics for this plant are given in table 2–2.

Table 2–2. Statistics for the Great Plains Synfuels Plant.

Daily Consumption	
Lignite (tons)	18,300
Power (kWh)	79,000
Water (million gallons)	6.6
Oxygen (tons)	3,100
Daily Production	
Natural gas (MMcf)	160
Phenol (tons)	48
Cresols (tons)	48
Ammonia (tons)	70
Krypton/Xenon (liters)	10,500
Sulphur (tons)	50
Nitrogen	various amounts

Source: Dakota Gasification Company brochure, 1996

SNG can be produced from any fossil fuel. Although coal has been of prime interest, there has been an increase in interest in using residual fuel and biofuels (wood).[34]

References

1. Woolen, I. "Central Asian Gas Crucial to Future Russian Gas Supply," *Oil & Gas Journal*, August 13, 2001, p. 61.

2. Al-Falih, K. A. "Saudi Arabia's Gas Sector: Its Role and Growth Opportunities," *Oil & Gas Journal*, June 21, 2004, p. 18.

3. Aitani, A. M. "Big Growth Ahead for Saudi Gas Utilization," *Oil & Gas Journal*, July 29, 2002, p. 20.

4. Poruban, S. "East Mediterranean Natural Gas Prospects, Projects Hit Pay Dirt," *Oil & Gas Journal*, May 20, 2002, p. 20; J. C. Dolson, et al. "Petroleum Potential of an Emerging Giant Gas Province, Nile Delta and Mediterranean Sea off Egypt," *Oil & Gas Journal*, May 20, 2002, p. 32.

5. Dittrick, P. "Independents, Other Companies Find Opportunity in Basins off West Africa," *Oil & Gas Journal*, March 4, 2002, p. 66; S. Shu, F. Christiano, and M. Harrison. "Analysis Points to Electric-Motor Drivers for Angola LNG," *Oil & Gas Journal*, October 7, 2002, p. 60.

6. Prates, J.-P. "Brazil's Energy Crisis Complicates Progress in Gas, Power Markets, but Outlook Brightening," *Oil & Gas Journal*, October 15, 2001, p. 77.

7. "Gas the Cornerstone of Venezuela's Energy Sector Investment Campaign," *Oil & Gas Journal*, April 24, 2000, p. 21.

8. Williams, B. "Trinidad and Tobago's Atlantic LNG Follows Initial Success with Aggressive Expansion Plans," *Oil & Gas Journal*, March 11, 2002, p. 22; and "Trinidad and Tobago Banking Its Future on Natural Gas, but Energy Policy Still Evolving," April 1, 2002, p. 22.

9. "Pluspetrol's Benito: Peru's Future Lies in Natural Gas," *Oil & Gas Journal*, November 25, 2002, p. 22; "NW Peru NGL-GTL Venture Has Exploration Component," *Oil & Gas Journal*, December 17, 2001, p. 34.

10. Scott, J. "TransCanada Growth Strategy Focused on N. American Gas, Power," *Oil & Gas Journal*, January 21, 2002, p. 18; "Canada's Huge Gas Potential Challenged by Cost, Pipeline Transport Issues," *Oil & Gas Journal*, January 20, 2003, p. 18; B. DeBaie. "Resource Base, Pipeline Networks Position Canadian Producers for Greater Share of U.S. Oil and Gas Demand," *Oil & Gas Journal*, June 28, 1999, p. 34.

11. Monteforte, R. "Gas Demand Growth Will Push Expansion in Mexican Transmission Infrastructure," *Oil & Gas Journal*, February 11, 2002, p. 70; M. P. Hoyt. "E&D Opportunities May Loom in Mexico's Natural Gas Industry," *Oil & Gas Journal,* November 12, 2001, p. 48.

12. Williams, B. "Route Controversy Heats Up," *Oil & Gas Journal*, August 6, 2001, p. 20.

13. Swain, E. J. "U.S. LNG Imports Poised to Play Long-Term Supply Role," *Oil & Gas Journal*, January 24, 2000, p. 38.

14. Fesharaki, F., and H. Vahidy. "India, Gas Business Comes of Age," *Hydrocarbon Asia*, November/December 2004, p. 6.

15. "India's Power Projects Drive Boom in LNG Import Scheme," *Oil & Gas Journal*, October 4, 1999, p. 25; "Indian LNG Projects Boom," *Oil & Gas Journal*, June 19, 2000, p. 62.

16. Vahidy, H., and F. Fesharaki. "Pakistan's Gas Discoveries Eliminate Import Need," *Oil & Gas Journal*, January 28, 2002, p. 24.

17. Shamsuddin, A. H. M., T. A. Brown, and M. Rickard. "Resource Studies Indicate Large Gas Potential in Bangladesh," *Oil & Gas Journal*, April 22, 2002, p. 48; ibid, "Bangladesh Gas Reserve, Resource Potential May Be Underestimated," April 29, 2002, p. 40 (extensive references); and F. Fesharaki. "Bangladesh Natural Gas Exports to India," *Oil & Gas Journal*, June 19, 2000, p. 20.

18. Prawiraatmadja, W. "Uncertainties Plague Indonesia's Gas Development Projects," *Oil & Gas Journal*, October 16, 2000, p. 34; S. L. Montgomery, et al. "Potential Giant Gas Reserves Await Development in Irian Jaya," *Oil & Gas Journal*, June 25, 2001, p. 40.

19. Fesharaki, F. "Fundamentals and Forecasts of Thailand's Natural Gas," *Hydrocarbon Asia*, March/April 2004, p. 12.

20. "Economic Woes, Sagging Demand Slow Thai Gas Pipeline Projects," *Oil & Gas Journal*, November 16, 1998, p. 27.

21. "Myanmar's Upstream Sector Hobbled by Pipeline Controversy, Poor E&D Results," *Oil & Gas Journal*, June 26, 2000, p. 24.

22. "First Commercial Sales Flow from Malampaya Gas Plant, Pipelines," *Oil & Gas Journal*, January 7, 2002, p. 58.

23. Omiya, M., et al. "Tight Schedule Prompts Staged Start up for Viet Nam's First Gas Plant," *Oil & Gas Journal*, September 25, 2000, p. 66.

24. Ellsworth, C., and R. Wang. "China's Natural Gas Industry Awakening, Poised for Growth," *Oil & Gas Journal*, July 5, 1999, p. 23.

25. Yun Long, W, "Current Status and Development Trend of China's Urban Gas Supply," *Hydrocarbon Asia*, November/December 2004, p. 10.

26. Hansen, R., and O. Olsvik. "Norwegian Methanol Plant Outlines Operations, Expansion Plans," *Oil & Gas Journal*, February 7, 2000, p. 46.

27. Mavor, M., T. Pratt, and R. DeBruyn. "Study Quantifies Powder River Coal Seam Properties," *Oil & Gas Journal*, April 26, 1999, p. 35; D. E. Gaddy. "Coalbed Methane Production Shows Wide Range of Variability," p. 41; C. D. Jenkins, et al. "Appraisal Drilling Focuses on Ordos Basin Coal Seams," p. 43.

28. Stevens, S. H., K. Sani, and S. Hardjosuwiryo. "Indonesia's 337 tcf CBM Resource a Low-Cost Alternative to Gas, LNG," *Oil & Gas Journal*, October 22, 2001, p. 40.

29. Gorody, A. W. "Coalbed Methane Production Faces Numerous Concerns," *Oil & Gas Journal*, July 23, 2001, p. 66; and "Base Line Studies Assess Regional Aquifers," *Oil & Gas Journal*, July 30, 2001, p. 51.

30. Makogan, Y. E., and S. A. Holditch. "Lab Work Clarifies Gas Hydrate Formation, Dissociation," *Oil & Gas Journal*, February 5, 2001, p. 43.

31. Makogan, Y. E., S. A. Holditch, and T. Y. Makogon. "Russian Field Illustrates Gas-Hydrate Production," *Oil & Gas Journal*, February 7, 2005, p. 43; "Blake Ridge Provides Data for Assessing Deepwater Gas Hydrates," *Oil & Gas Journal*, February 14, 2005, p. 45.

32. Rach, N. R. "Japan Undertakes Ambitious Hydrate Drilling Program." *Oil & Gas Journal*, February 9, 2004, p. 37; and H. Takahashi, T. Yonezawa, and E. Fercho. "Overview of the Mallik Gas-Hydrate Production Research Well," Offshore Technology Conference, Houston, May 5–8, 2003 (OTC Paper 15124), abridged in *Journal of Petroleum Technology*, April 2004, p. 53.

33. Shanley, K. W., J. Robinson, and R. M. Cluff. "Tight-Gas Myths, Realities Have Strong Implications for Resource Estimation, Policymaking, Operating Strategies," *Oil & Gas Journal*, August 2, 2004, p. 24.

34. Nahas, N. C. "Catalytic Methane Synthesis Can Extend Hydrocarbon Supply," *Oil & Gas Journal*, October 4, 2004, p. 18.

3

Gas Composition

The composition of natural gas influences its use. Some gases have significant issues relating to their composition that have a profound influence on their utility. This raises the cost of producing a gas acceptable to a downstream user.

When gas comes to the surface, it contains many components that need to be extracted before it can be used. The components are extracted in different unit operations, the choice and size of which are dependent upon the raw gas composition and the amount of component being extracted. Taken together, the different unit operations are referred to as the *gas plant*. There are many choices of the design of the gas plant, and all gas plants are unique. For any gas and downstream application, there are usually several viable technical and economic solutions. An analysis of gas plant design is beyond the scope of this book, which will only consider the principal issues. For additional information, the reader is referred to Newman, who details 28 approaches to gas plant design, and the "Gas Processes" editions of *Hydrocarbon Processing*. The reader also is referred to the *Oil & Gas Journal,* which regularly publishes articles on gas plant design.[1, 2]

Many gas pipeline systems have open access arrangements. However, gas quality is controlled by rather tight specifications. For the most part, these specifications concern protecting the integrity of the pipeline [e.g., carbon dioxide (CO_2) content] or the *Wobbe Index* for users downstream.

The problem for a company with a new gas development that wishes to access an existing pipeline system is that alteration to the raw gas composition will almost always be required. Of course, if the gas and pipeline represent a new development, then the specification can be optimized to minimize the processing costs. In some cases, a tight specification based on an original owner's/supplier's gas stream can be used to shut out other users from the system and maintain a monopoly control.

This chapter will examine issues relating to composition and upgrading of gas to compositions of better use to utilities and downstream users. An overview of the main technologies and approximate cost for upgrading gas from published data are given.

Variation in Gas Composition: Nonhydrocarbons

There is considerable variation in natural gas composition worldwide, which will influence the utility of a particular field for a particular end use.

In many instances the comprehensive analysis of a given gas field is unavailable, especially in terms of trace components such as helium. This is because most exploration is conducted with fuel usage as a priority. Hence minor components that do not contribute to calorific value are often ignored.

Before the composition of natural gas in terms of the principal hydrocarbons is discussed, some pertinent components, a number of which can be extracted and marketed, are reviewed.

Water

Raw natural gas is saturated with water. In addition, slugs of water and oil flow to the surface with the gas. Appropriately sized separation vessels and long, wide diameter tubes known as *slug catchers* remove these. The water content of gas is further reduced to suppress the likelihood of condensation in pipelines. This might be undertaken for more than one reason, such as the following:

1. Offshore gas processing. Before gas can be recovered and sent to shore-based gas plants, excess water is removed at a platform treatment facility. Further drying reduces the likelihood of the condensation of water in the transmission line and the formation of ice or natural gas hydrates. The latter is a particular problem with gas at high pressure flowing through pipelines below 20°C, which is the case for most undersea pipelines. To increase the suppression of hydrates, glycol or methanol is sometimes added.[3]

2. Production of pipeline-quality gas. Pipeline-quality gas requires the water content of the natural gas to be reduced to prevent condensation in the transmission system. The *dew point,* or the temperature at which condensation occurs, is reduced to very low levels (typically –30°C). This helps to control corrosion in the transmission system.

In addition, some downstream processing operations require gas to be rigorously dried to prevent ice formation in cryogenic operations.

Technology. There are two general approaches for water absorption using drying sieves or alumina and absorption using glycol or methanol.[4] In sieve drying, the wet gas is contacted with a bed of absorbent. Heating regenerates the absorbent bed. In the glycol drying process, wet gas is contacted with triethylene glycol (TEG). The TEG is regenerated by using a dry stripping gas, or more commonly, by a reboiler. An issue is that TEG also absorbs heavier hydrocarbons that are subsequently vented to the atmosphere after the reboiler. Kirchgessner and others have described an advanced process that minimizes these emissions.[5]

Absorbent systems are used widely in large gas plants, and glycol-drying systems are often found on platforms.

Nitrogen

Nitrogen is found in small quantities in most natural gas deposits. At a level of less than 2%, which is typical, it does not have a major influence on the gas properties.

However, some gases contain large concentrations of nitrogen. At a level of 10% and higher, there may be a significant and unacceptable drop in the calorific value, and treatment may be required to remove it.

Nitrogen is difficult to remove. The conventional method is by liquefaction, where all of the other components are condensed, and methane and nitrogen are separated by distillation. The required components for sales gas are recombined. This method is very capital intensive.[6] Workers at the Gas Research Institute have claimed to have optimized processes for smaller gas fields.[7]

A variation of cryogenic separation is described by Advanced Extraction Technologies. This process utilizes a refrigerated solvent to separate the methane from the nitrogen by a series of flash vessels.[8] The method claims lower capital and operating costs than conventional cryogenic methods.

An alternative approach is to extract the hydrocarbons by solvent extraction or by pressure swing absorption (PSA). These methods are usually performed in multiple stages. The cost of removal is relatively high, typically $0.50–$0.80/Mcf of sales gas.[9]

Helium

Helium is found in small quantities in almost all natural gas deposits. The source of helium is radioactive decay from either minerals surrounding the deposit or from deeper in the earth where it slowly permeates to the surface. During migration, minor quantities become trapped in subsurface petroleum reservoirs, with older formations often containing the highest concentrations. High concentrations are found when gas reservoirs are located over buried granite.

For concentrations >0.3%, it is commercially viable to extract helium. Some natural gas in the United States contains several percent helium. Some fields also contain large quantities of nitrogen. The U.S. fields are the predominant source of the world's commercial helium. Extraction is by exhaustive cryogenic liquefaction of all components other than helium.

Oxygen

Oxygen is not normally determined in the analysis of gas, and when present is usually in fractions of 1% or less. Oxygen can also enter the gas by air entrainment in the normal operations of equipment and in the analytical sampling tools. This makes measurement and control of oxygen content difficult. Up until recently, small amounts of oxygen were not considered a problem, since pipeline specification is typically 0.2% maximum. However, recently there has been some concern regarding trace amounts of oxygen leading to degradation of amine in acid gas plants forming stable salts.[10]

Mercury

Many natural gases contain trace quantities of mercury (typically in the order of micrograms per cubic meter). Mercury is particularly deleterious to cryogenic processes because it corrodes the low temperature aluminium alloy heat exchangers. Mercury discharge from production facilities (in the produced water) has also been a problem for some fields. Unocal has stated that the mercury in the water discharge from its fields in the Gulf of Thailand would be in the order of 300 kg/year if treatment were not used.[11]

Mercury has to be removed before LNG manufacture, and it would be prudent for processing or conversion involving high activity catalysts. This is easily achieved at a relatively low cost using sulfur-impregnated carbon.[12]

Sulfur

Sulfur is present in high concentrations in many gas fields in North and South America, Europe, and the Middle East. However, in other areas (e.g., Australia), there is negligible sulfur present in the gas. Analytical figures for sulfur often represent the detection limit of the analytical method in use. Sulfur has to be removed from the natural gas for nearly all downstream uses.

Sulfur is present as hydrogen sulfide, carbonyl sulfide, or mercaptans. Mercaptans and carbonyl sulfide are not normally removed in older acid gas plants. When present, they are usually in small amounts. These small amounts of sulfur can be removed by absorption at ambient temperatures on activated charcoal or with molecular sieves. Often multiple beds are used to remove trace quantities of different species.[13] The more modern acid gas technologies use a mixed solvent that can remove carbonyl sulfide and mercaptans as well as hydrogen sulfide and carbon dioxide.[14]

If the natural gas is not treated and the components of natural gas are separated, carbonyl sulfide [boiling point (b.p.) of –50°C] often contaminates the propane fraction (b.p. of –42.1°C). There are several absorbent and regenerable processes for carbonyl sulfide removal from propane streams.[15]

Hydrogen sulfide. There are many proprietary variants on the available technology.[16] If hydrogen sulfide is the only contaminant present, then it may be possible to remove it with a bed of zinc oxide, by use of iron oxide, or by liquid absorption.[17, 18]

Hydrogen sulfide and carbon dioxide are acidic and are generally removed together in acid gas plants (fig. 3–1). In these plants, the gas is contacted by an alkaline medium, often a solution containing an organic amine, alkanolamine, or amine mixture that forms a salt with the acid gas. The spent amine is then regenerated, most often by heating, which breaks the salt, regenerating the amine and liberating the acid gas.

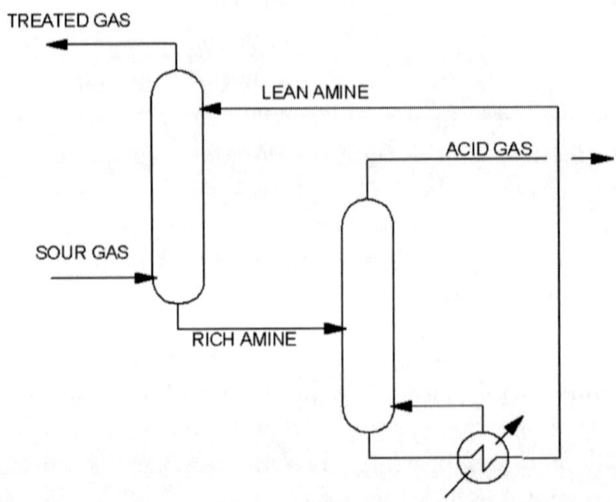

Fig. 3–1. Acid gas plant basic layout

In order to lower the absorption of higher hydrocarbons, sorbents based on N-formylmorpholine (NFM) have been developed.[19]

Higher concentrations of hydrogen sulfide in the off gas are treated using the Claus process, in which hydrogen sulfide is reacted with sulfur dioxide to produce elemental sulfur. (The Claus process is an old process and nowadays exists in many proprietary variants, which are described in the referenced texts.)

Claus and alternatives to the Claus process. Claus plants are used to reduce the hydrogen sulfide extracted in acid gas plants and to produce sulfur. The increasing concern with environmental emissions of sulfurous gases (*acid rain* emissions) has resulted in the application of additional gas treatments to the Claus plant.

The Claus and related processes are the largest source of elemental sulfur for fertilizer and industrial uses. The Claus process is widely used in *sweetening* the natural gas in Canada, where some producing wells contain as much as 50% hydrogen sulfide. The elemental sulfur is collected and exported from Vancouver in large volumes.

As previously stated, there are many approaches to the treatment of sour natural gases and alternatives to the treatment of the acid gases that are produced as off gases to the gas plant. The Claus process (developed in 1880) is one of the oldest. In this process, some of the hydrogen sulfide in the acid gas is burned to form sulfur dioxide. The remainder and the sulfur dioxide next pass to a reactor containing a catalyst, where sulfur is formed and then condensed and removed as a liquid.

The chemical processes are:

$$2H_2S + 3O_2 = 2SO_2 + 2H_2O \tag{3.1}$$

and

$$2H_2S + SO_2 = 3S + 2H_2O \tag{3.2}$$

About 95% of the hydrogen sulfide is removed. Additional units or separate Claus off-gas units are required to eliminate the remainder.

Over the years the process has been improved, and it is now available in several proprietary forms.

Other approaches to handling acid or sour gases are also used. For instance, Jones and others describe the treatment of the sour off gases (14 MMscfd containing 4% H_2S, 94% CO_2, and 2% H_2O) from an amine treatment unit.[20] This is accomplished by a solvent extraction process (Flexisorb) in which carbon dioxide is separated from the hydrogen sulfide. The carbon dioxide (and any hydrocarbons that may have dissolved) are passed to a thermal incinerator before venting. The hydrogen sulfide gas (1 MMscfd > 95% H_2S) is compressed and reinjected via a gas well into the reservoir where the gas originated.

Odorant. Sulfur in the form of thiophene or a thiophene derivative (such as tetrahydrothiophene, THT) is added to gas as an odorant (about 7 mg/m^3). This occurs before reticulation to domestic consumers, and hence it can be present in pipeline gas. Odorant levels of sulfur would be deleterious to some conversion processes, e.g., steam reforming, and would have to be removed before use.

Carbon dioxide

Some gas fields contain high levels of carbon dioxide. Carbon dioxide has no calorific value and as a pure substance has limited industrial use. Such uses are mainly as an inert gas or carbonating agent, or in urea production and in certain enhanced oil recovery (EOR) schemes. The principal source of pure carbon dioxide in the industrialized world is as a by-product in hydrogen and ammonia manufacture. Substantial quantities of the gas are produced, much of which is discharged to atmosphere. However, where demand is high, for instance in EOR schemes, gas fields can be developed for their carbon dioxide content.

Because of its inert nature, it can be a hindrance in the use of some natural gases by lowering the calorific value. It can also promote the corrosion of pipelines. However, if the gas contains substantial amounts of carbon dioxide, the cost of removal becomes significant.

If LPG is extracted from natural gas, the carbon dioxide content is generally reduced to about 2% in the process, thus raising the calorific value of the product. For certain uses, especially methanol production, the presence of some carbon dioxide can be an advantage. It can help bring into balance the hydrogen/carbon (H/C) ratio of the feedstock and product. Thus fields with poor marketability as fuel due to carbon dioxide content may be eminently suitable for methanol production.

The present concern with greenhouse gas emissions to the atmosphere has resulted in considerable interest in the extraction and sequestration of carbon dioxide. This concern has deterred the timely development of some large gas fields that would otherwise be suitable for LNG production. Such fields include the Gorgon field in Australia (10% CO_2) and the Natuna Field in Indonesia (70% CO_2).

There are a wide variety of approaches to the removal of carbon dioxide.[21] One common approach is through the acid gas plant.[22] Provided the presence of sulfur does not present a problem, carbon dioxide can also be removed using membrane separators and molecular sieve separators. These technologies have the advantage of being small and compact. They are particularly useful for smaller gas field developments or offshore operations where size is at a premium.

Membrane separators. Membrane technology works on the principle that different gases diffuse at different rates through different types of polymer membranes. Carbon dioxide, water, and hydrogen are particularly fast and can be easily separated from slow diffusing molecules such as hydrocarbons or nitrogen. There are a broad number of applications of membrane technology and many proprietary technologies.[23]

The basic method is illustrated in figure 3–2. For carbon dioxide separation, gas is passed at pressure through one side of the membrane, and carbon dioxide and water preferentially diffuse through the membrane and are exhausted from the low-pressure side.

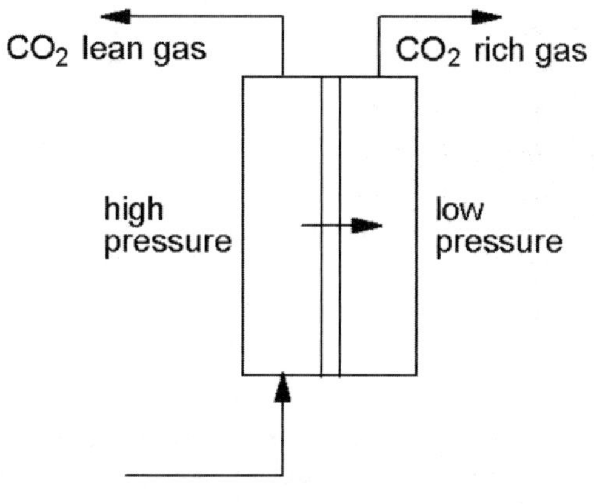

Fig. 3–2. Membrane separation of carbon dioxide

Single stage systems (as illustrated) are not very effective in the sense that a certain amount of hydrocarbon gas also passes through the membrane. To overcome this problem, membrane systems are normally operated with multiple separators using interstage compression as necessary. Membrane systems are claimed to offer significant cost advantages over acid gas plant technology.[24]

Molecular sieve separators. In molecular sieve separators, a sieve preferentially absorbs carbon dioxide and water. When the sieve is saturated, it is taken offline, and the gases are desorbed by depressurizing. Any methane desorbing can be separated (differential desorption) and recompressed. The carbon dioxide is obtained as a low-pressure gas.[25]

They are particularly useful in removing carbon dioxide from coalbed methane. However, some issues occur in using these systems in the presence of heavier hydrocarbons.

Although membrane and molecular sieve separators are smaller than amine absorbers, they are not as efficient. Further, there is a higher loss of methane through the system, which is ultimately discharged with the carbon dioxide.

Variation in Hydrocarbon Content

Liquid products produced from gas come under a variety of names. *Natural gas liquids* (NGL) is a generic term for all condensed products. The C_{5+} fraction (boiling point > 30°C) is often referred to as condensate, or sometimes, especially in the United States, as *natural gasoline*. The C_3 and C_4 fraction is liquefied petroleum gas.

When considering the higher hydrocarbons present in natural gas, it is probably best to recall that a continuum exists in oil and gas reservoirs. This ranges from almost pure methane to heavy petroleum oils and waxes. In general, hydrocarbon deposits do not span to the extremes. Higher hydrocarbon-free natural gas is not commonly accounted, although there are some very large gas deposits that are comprised almost entirely of methane. Thus many oil reserves have considerable quantities of associated gas, and most gas reserves have associated light oil (condensate) deposits.

Depending upon quantity and location, associated gas (i.e., from a producing oil well containing gas) is separated at the wellhead and transported. If transportation is not possible, current practice is to flare or reinject the associated gas. Large quantities of gas can be wasted in this manner.

Associated oil, or condensate, in primary gas-producing locations can be an important economic factor in the development of a particular field. In the planned development of remote gas fields, the initial venture might first be the recovery of condensate with gas reinjection. This would generate early cash flow before developing the natural gas.

Liquid condensate is easily separated and transported. Its availability, market, and end use are beyond the scope of this book.

Table 3–1 illustrates the variation in composition of some developed natural gas reserves of Australia and Asia, together with a typical analysis of U.S. dry natural gas. These have been chosen to illustrate the principal issues. The principal features are illustrated in figure 3–3.

Table 3–1. Analysis of some natural gases from Australia, Indonesia and US (Vol %)

	C1	C2	C3	i-C4	n-C4	i-C5	n-C5	C6+	N_2	CO_2
Barracouta (Aus)	85.4	5.9	3.1	1.2	0.6	0.8	0.02	0.6	1.8	0.8
Marlin (Aus)	80.2	7.0	4.4	0.8	1.5	0.6	0.6	2.2	0.7	2.0
Kipper (Aus)	77.7	6.1	2.7	0.5	0.8	0.3	0.3	2.0	0.3	9.5
Cooper Basin (Aus)	72.9	5.3	1.9	0.2	0.5	0.1	0.2	0.9	1.0	17
Goodwyn (Aus)	81.2	7.9	3.8	0.6	1.3	0.4	0.5	1.2	1.7	1.4
Palm Valley (Aus)[a]	87.8	8	1.2	0.1	0.3	0.1	0.1	0.1	2.2	0.4
Mereenie (Aus)[a]	71.3	13	4.6	0.5	1.3	0.3	0.3	0.1	8.5	0.1
Woodada (Aus)	91.1	1.1	0.4	0.1	0.1			0.1	3.1	4.2
Scott Reefa (Aus)	80.3	5.3	2.3		0.9				0.3	11
Vic Pipeline (Aus)[b]	90.7	5.9	0.6						0.9	1.8
Arun (Ind)[a, c, d]	74.9	5.6	2.3	0.5	0.6	0.2	0.2	0.4	0.4	15
Sengeti (Ind)[d]	51.6	14.8	15.2	3.5	5.3	2.1	1.4	0.1		6.2
U.S. Salt Lake (UT)	95.0	0.8	0.2						0.4	3.6
U.S. Webb (Tx)	89.4	6.0	2.2	1.1	0.7				0.1	0.6

Notes: (a) After condensate removal (b) Gas and Fuel Corp. ex. Longford
(c) Also contains 100 ppm H_2S and 0.11mg/m3 mercury (d) Indonesia (Sumatra)

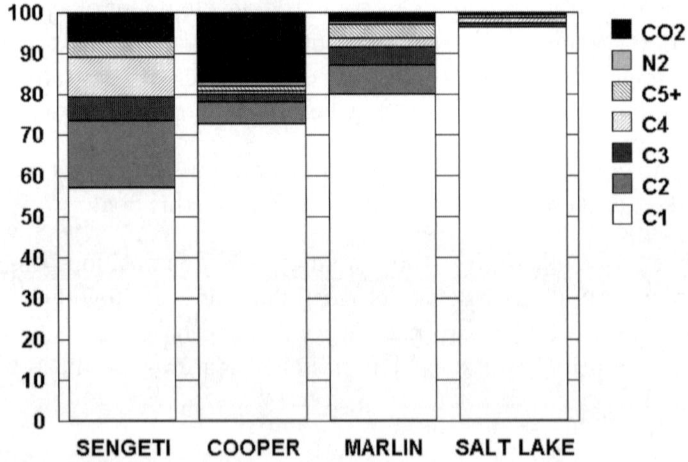

Fig. 3–3. Typical wellhead gas compositions

The principal component is methane, typically 80% by volume. Of the higher hydrocarbons, there is a complete absence of unsaturated compounds (olefins), but where possible, simple branching is present. Carbon dioxide is present in levels from about 1% and higher. Some gases contain high levels of carbon dioxide (10%). In the Australian gases, the nitrogen level is low.

Note that the conventional method for reporting gas composition is in terms of volume percent. This approximates to mole percent, which is also widely used. The gas molecular weights are approximately 18. If expressed on a weight basis, the figures for methane (molecular weight 16) fall with a concomitant rise in the figures for higher molecular weight components. This is of particular concern for carbon dioxide (molecular weight 44), which when expressed on a weight basis increases the figure by approximately 50%. A weight-based analysis is important in estimating potential greenhouse (carbon dioxide) emissions.

The calorific value of the major hydrocarbon component, methane, in volume terms is relatively low. The higher heating value (HHV) of methane is about 1,010 BTU/scf (37.56 MJ/m^3). For the higher hydrocarbons, it is considerably higher. Ethane is 1,769 BTU/scf (65.8 MJ/m^3), and propane is 2,517 BTU/scf (93.65 MJ/m^3). Thus the presence of the higher hydrocarbons significantly improves the value of the gas as a fuel. A consequence of this is that the higher hydrocarbons, especially ethane, are often left in the gas in order to lift the calorific value in volume terms. Such action overlooks the value of the higher fractions as petrochemical feedstock.

General Approaches to Gas Treatment

Before discussing the detail of hydrocarbon removal, a general approach to gas treatment prior to removal of heavier hydrocarbon liquids is given. These generalized flow schemes draw on the data for individual unit operations discussed in the section above.

Low-sulfur, low-carbon dioxide gas

Figure 3–4 illustrates a typical flow sheet for gas treatment of a low-sulfur, low- carbon dioxide gas (sometimes referred to as a sweet gas).

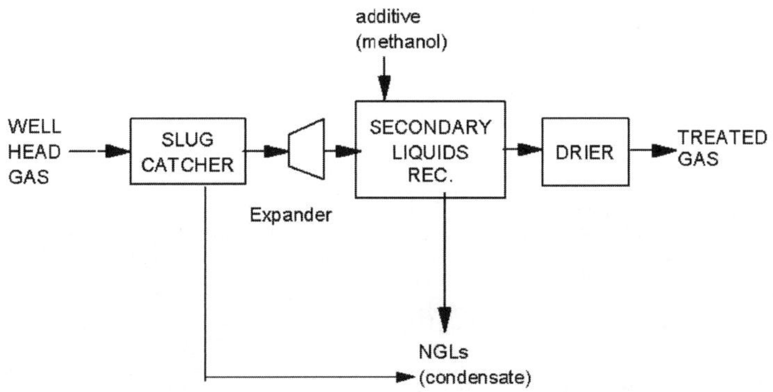

Fig. 3–4. Gas treatment for low-sulfur, low-carbon dioxide gas

Such a scheme is simple, and the capital cost is low. This type of scheme can be used on low-sulfur and low-carbon dioxide gases where gas is supplied to a limited number of end users, and the specification can be somewhat relaxed.

High-sulfur, high-carbon dioxide gas

This type of gas is sometimes referred to as a sour or acid gas. For this case, gas processing involves many steps, and there are several alternatives to the order of unit operations. There are a large number of proprietary technologies available. An absorber is central to the treatment for the majority of technologies available. This is where the acidic hydrogen sulfide and carbon dioxide are absorbed into a solvent. As discussed above, there are many variations on the types of solvent in use. Figure 3–5 illustrates a typical flow sheet for treatment of a gas with a high sulfur and/or high carbon dioxide content.

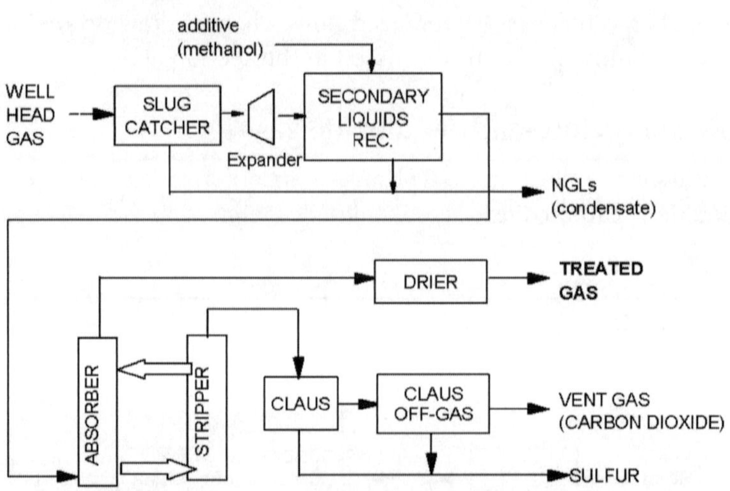

Fig. 3–5. Gas treatment for high-sulfur, high-carbon dioxide gas

The flow sheet shown has the following steps:

- Pretreatment and condensate removal as previously described (fig. 3–4).
- An acid gas treatment plant to remove hydrogen sulfide and carbon dioxide. This comprises two towers. Gas enters the bottom of the first tower, where it contacts a solvent that strips the acid gases out of the system. The solvent passes to the stripping tower, where it is recovered by boiling (or sometimes by pressure swing).

- The treated gas passes out of the top of the first tower and is dried before transmission. The off gas from the stripping tower is passed to a Claus plant where several steps separate the sulfur. Emissions regulations may force the adoption of additional unit operations to remove the last vestiges of sulfur from the Claus off gas. The residual vent gases are then principally carbon dioxide.

Such a gas treatment is considerably more extensive and expensive than the simple scheme discussed for low-sulfur gas.

Gas Treatment to Remove LPG and Ethane

In order to remove LPG (propane and butane) from the wet gas stream, current processes require the gas stream to be chilled to –20°C or below. (Note that the term *wet gas* in this context means that the gas contains condensable hydrocarbons.) For reasons described above, this requires the removal of water and carbon dioxide from the gas stream. There are two main processes: turbo expansion and refrigerated solvent absorption.

LPG removal by turbo expansion

The basic flow for a turbo-expander scheme is illustrated in figure 3–6. This represents the simplest flow diagram, which can be quite complex if ethane is to be extracted.[26]

In the turbo-expander method, wet gas (water dry) is compressed, typically to 100 bar. The heat of compression is removed in an interstage cooler. The gas is then passed to an expander, which is coupled to the compressor in order to recover some of the required shaft power. This causes the gas stream to cool to below the liquefaction point of the LPG.

Condensed LPG and gas is passed to a flash vessel, which separates the dry gas from the liquids. The liquids are passed to a distillation column, where LPG and condensate are separated. Propane and butane can be separated in an additional column.

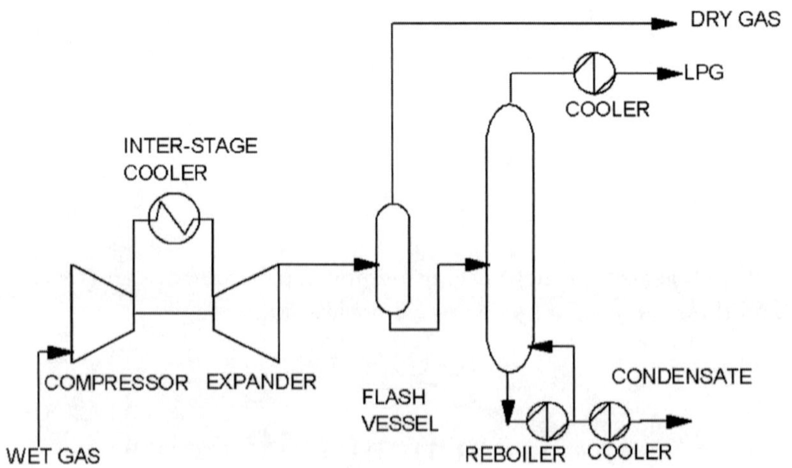

Fig 3–6. LPG extraction by turbo expander

Over time, turbo-expander systems have improved in efficiency and can be used to extract ethane by inclusion of gas-to-gas heat recovery systems.[27] These are variously described as cryogenic systems or cold boxes and are similar in operation to the cryogenic units used for the production of LNG. The use of cold boxes permits cooling of the gas before it reaches the turbo expander, and hence results in an overall colder operation, which is illustrated in figure 3–7.

Inlet gas enters the first cold box. Here the gas is chilled, and separated liquids are passed to a large de-methanizer column. A second cold box repeats the process, after which the cold gas is expanded to condense the remaining liquids. The cold gas is now passed to the top of the de-methanizer, where it is used to cool the incoming gas in the cold boxes.

Such systems can recover up to about 80% of the ethane present. Addition of further cooling to the top of the de-methanizer can achieve more than 90% ethane recovery.[28]

One point of note is that the use of cold box technology requires the removal of mercury from the gas streams. A fire at Santos' Moomba facility in Australia in early 2004 was thought to be due to mercury attack on equipment.

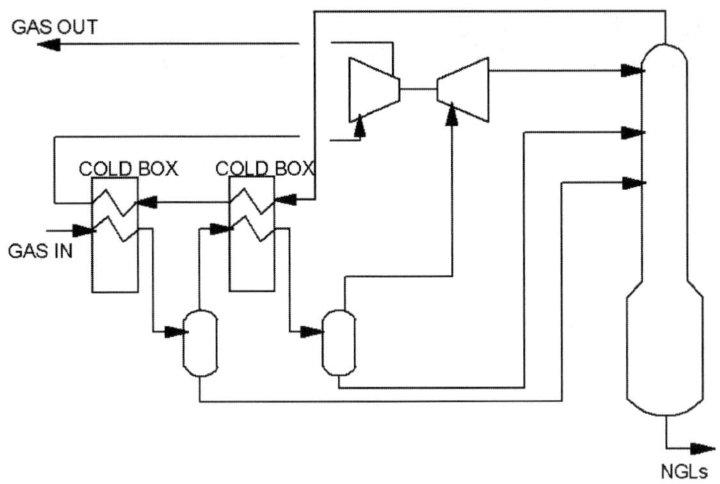

Fig. 3–7. Ethane and LPG extraction using cold boxes

Straddle plants. One advantage of the turbo-expander method for separating LPG from natural gas is that it allows the use of gas pipelines to transport the LPG. LPG is costly to store and transport, as it requires pressurized or cryogenic vessels. By using gas pipelines, the lower cost transport economics of pipeline gas can be used.

In the straddle plant option, LPG is left in the sales gas at the gas plant. The much-larger volume of methane dilutes the LPG, and the gas (including the LPG) meets the pipeline dew-point specification. The mixture is then piped over several hundred kilometers to the straddle plant. This uses a turbo expander to separate LPG from the gas, maintaining the residual gas within the heating value specification.

There are several such operations in Canada and Australia, which have been described by Hawkins.[29] Data for the large Canadian plants, which process 9.8 Bcf/d of gas, are illustrated in table 3–2. This shows that although the concentration of NGL in the gas is low, large volumes are recoverable, given the large gas flows available.

Table 3–2. Data for Empress straddle-plants in Alberta

Component	Vol%	gal/Mcf	Recovery (%)	Recovered gal/Mcf
Methane	92.02			
Ethane	4.25	1.1	70.0	0.77
Propane	1.00	0.3	90.0	0.27
i-Butane	0.10	0.0	99.5	0.00
n-Butane	0.20	0.1	99.9	0.10
n-Pentane	0.05		100.0	0.00
n-Hexane	0.02		100.0	0.00
n-Heptane	0.00		100.0	0.00
n-Octane	0.00		100.0	0.00
CO_2	0.31			
N_2	2.06			
Total	**100.01**	**1.5**		**1.14**

From the data in table 3–2, the properties of the gas can be estimated before and after the straddle plant, as shown in table 3–3. This shows that although the gas loses some of its heating value, it remains within a specification of 1,000–1,050 BTU/scf.

Table 3–3. Typical properties for gas streams across straddle-plants

	Input	Output
BTU/ft^3 gross	1047.3	1000.0
BTU/ft^3 net	945.8	901.6
HHV MJ/cm	39.1	37.3
LHV MJ/cm	35.3	33.7
Mol Wt.	17.4	16.6

Refrigerated absorption plants

Before the advent of turbo-expander plants in the early 1970s, the preferred method for removal of LPG materials from the gas stream was by absorption in a suitable solvent. To increase the absorption efficiencies, especially for the recovery of ethane, this technology was developed by applying refrigerated solvent to the gas stream.

The absorption plants use a hydrocarbon solvent, similar to kerosene in boiling range. This is chilled to about –25°C and used to absorb the required components. Because of the low temperatures, gas entering the system has to be water dry and low in carbon dioxide. Consequently, these are removed in upstream operations.

Absorption plants are comprised of three parts:

1. An absorber section
2. A section to remove dissolved gas, which is returned to the sales gas stream
3. A distillation unit that expels the absorbed components and regenerates the solvent

Solvent that is free of absorbed components is referred to as *lean oil*, while solvent containing absorbed components is referred to as *rich oil*. The main flows are illustrated in figure 3–8.

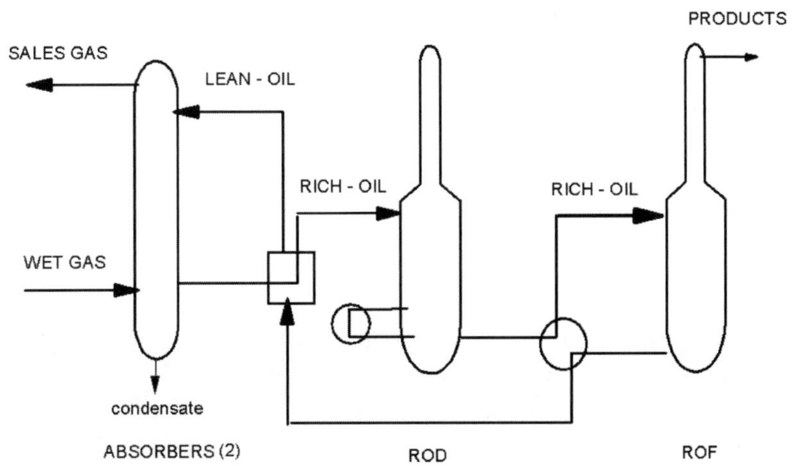

Fig. 3–8. Refrigeration absorption plants—main flows

Water-dried and acid-free gas (free of carbon dioxide and hydrogen sulfide), i.e., wet gas, is chilled to about –35°C and enters the bottom of the absorber tower. (There are usually two absorber towers.) Condensate separated in the chiller leaves the bottom of the tower, and the gas rises against a chilled falling solvent that has entered the top of the tower (lean oil). The solvent absorbs the heavier constituents, while the lighter sales gas rises to the top of the absorber and exits the top of the tower. The now-rich oil is collected on an absorber tray above the gas entry point and passes via heat exchangers to an ROD column. (The ROD column is variously described as a rich oil de-ethanizer when ethane and methane are to be expelled or a rich oil de-methanizer when methane alone is to be expelled.)

The ROD has the duty to remove any sales gas that may have dissolved and to return this to the sales gas stream. The rich oil enters near the top of the column and falls against warmed (about 50°C to 60°C) rich oil circulating through an exchanger. The degassed oil leaves the bottom of the tower, and the recovered gas leaves the top of the tower. This unit can be operated in two modes. If ethane is not a required product, the rich oil is heated sufficiently to expel ethane along with methane from the top of the tower. If ethane is to be extracted, the ROD is warmed to expel mainly methane.

The rich oil passes to the rich oil fractionator (ROF), where the solvent is boiled, regenerating the lean oil and expelling the LPG (and ethane) from the top of the tower.

As the fluids pass from the absorber to the ROF, the temperature rises from about –35°C to about 200°C (the b.p. of kerosene). This requires a considerable amount of heat exchanger capacity. Further, the pressure progressively falls from about 100 atm in the absorber to about 50 atm in the ROD to less than 10 atm in the ROF. This requires the lean oil stream to be pumped against this pressure rise.

The refrigerated absorber technology is complex and manpower intensive compared to the turbo-expander technology that has replaced it. However, where it still exists, it is particularly useful for recovering ethane, which is more difficult to extract in turbo-expander plants.

Gas treatment for LNG manufacture

For LNG manufacture, the gas has to be extensively treated in order to remove all of these products that are likely to freeze or cause damage to the liquefaction train. The flow sheet comprises the above unit operations plus mercury removal to produce the final gas. The cost of such gas treatment is high. LNG production is considered in more detail in chapter 11.

Offshore gas processing

Most offshore platforms for the production of oil and gas have some gas treatment facilities. When oil production is the main emphasis, the amount of process plant and storage can be quite complex and can vary with location.

For gas processing, few platforms have gas treatment facilities other than dehydration. Most pass the gas to an onshore facility for full processing to specification gas. However, platforms with dehydration and dew-point control to produce sales gas have been reported in Indonesia.[30]

Figure 3–9 illustrates a typical offshore gas treatment. Oil and gas are separated in three separators (not shown) operating at high, medium, and low pressures. The gas from these separators enters the gas treatment train at the points shown. The gas is recompressed and cooled before entering the glycol tower, where it is dried using an ethylene glycol derivative, usually triethylene glycol (TEG). Final compression and knock out of liquids occurs before the gas is exported to a shore-based processing facility.

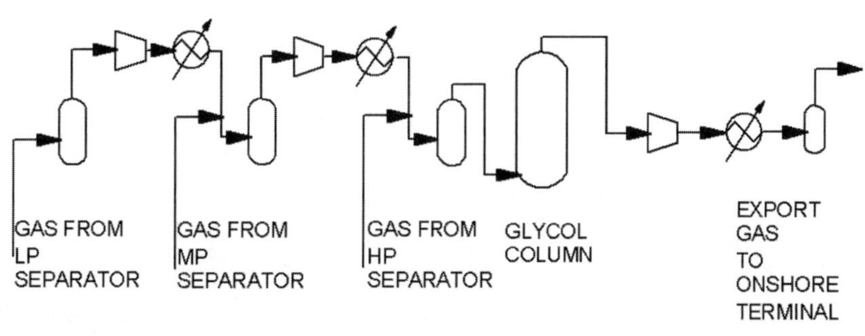

Fig. 3–9. Offshore gas processing

Gas Specifications

Across the world, natural gas is produced to a range of specifications.[31] Regulatory authorities ultimately sanction these specifications, with input from the gas industry as a whole. This includes the distribution companies and downstream users, including appliance manufacturers, the major pipeline operators, and companies wishing to supply gas. Because each of the groups approaches the issue from differing standpoints, they impose different specifications. The usual result is that the access to a given network is determined by the narrowest specification.

There are several key parameters that must be ensured:

- **The gas must be nontoxic and noncorrosive.** This involves the regulation of sulfur compounds, mercury, and radioactivity. Note that in many parts of the world, natural gas replaced coal gas, which is highly toxic and corrosive.
- **The formation of liquids or solids in the transmission and distribution system must be prevented.** This sets limits to the water content (which can form hydrates at low temperatures in high-pressure transmission systems) and hydrocarbon liquids.
- **It must permit interchange with gas from different suppliers.** This sets limits to the heating value, the Wobbe Index, and other combustion related matters. The Wobbe Index is a function of the heating value and the specific gravity. Setting the Wobbe Index and the heating value essentially limits the amount of carbon dioxide, which has a high specific gravity.

Over the years, three types of natural gas networks have emerged:

- In Asia, the gas is generally distributed as rich gas, with a relatively high HHV of typically 43 MJ/m3 (1,090 BTU/scf) and a concomitantly high Wobbe Index.
- By contrast, in the United States and the United Kingdom, the gas is lean, with an HHV of 42 MJ/m3 or less (1,065 BTU/scf).

- In continental Europe, the HHV ranges widely (39–46 MJ/m3; 990–1,160 BTU/scf). This permits import of gas from a variety of sources, ranging from lean gas from the Groningen field in the Netherlands to richer gas imported from North Africa, Russia, or Norway.

One of the problems for the gas industry, in particular the LNG industry, is that the gas specifications do not overlap, and gases meeting the lean gas standards in the United States and the United Kingdom will not meet the high calorific value gas required in Asia.

Another issue occurs in the development of alternative gas sources, particularly coalbed methane and natural gas hydrates. These alternative sources produce a very lean gas, with, except for methane, few other components. Methane has an HHV of about 38 MJ/m^3, so methane-rich gases have relatively easy access into the lean gas networks. However, in order to access a system with a high HHV specification, LPG has to be added. The cost of this increases the frequently already higher production costs for the alternatives and can shut them out of the distribution system.

Table 3–4 gives pertinent details of gas specifications from different countries.

Table 3–4. Some gas specifications

	UNITS	France	UK	Germany Group L	Germany Group H	USA (a)	New Zealand	Mexico	Australia (b)
Pressure (ref.)	bar	1.013	1.013	1.013	1.013	1.013	1.013	0.986	
Temp. (ref.)	C	0	15	0	0	15.15	15	20	
HHV (max.)	MJ/cm	41.56				38.46		41.53	
HHV (min.)	MJ/cm	34.74						36.03	37.3
LHV (max.)	MJ/cm	34.09							
LHV (min.)	MJ/cm	30.84							
H_2S (max.)	mg/cm	13.53	5.8	4.51	4.51		4.76	6.1	
H_2S (<8h)	mg/cm	13.31	4.76						
H_2S (8 day average)	mg/cm	6.61							
Sulphur (max.)	mg/cm	135.27	47.57	108.22	108.22		47.57	150	10
Water Dew Point	C	-5							
Humidity	mg/cm					118	100	112	58
Wobbe Index (max)	MJ/cm		48.91	42.2	50.97		49.47	50.6	51
Wobbe Index (min)	MJ/cm		44.9	37.8	41.56	47.37	43.76	45.8	47.3
H_2/C						3.92			
Specific gravity						0.6			
CO_2 (max)									
$CO_2 + N_2$ (max)			0.48			2.6		5	6
Incomplete Soot index			0.6						
Oxygen (max)			0.2				0.1	0.5	
Hydro-carbon dew point	C						2	-2	0
Bq/cm (max.)									600

Notes (a) averages; (b) Danbury-Bunbury pipeline—no LPG. From R. Monteforte, L. M. Damian, and M. Espinal. "Mexico's CRE Issues New Gas Quality Specs," Oil & Gas Journal, July 5, 2004, p. 62.

The table illustrates that different regulatory authorities focus on different parameters. Note that the measurement standard (temperature and pressure) differs from country to country.

Gas Production Costs: Order of Magnitude Estimates

This section provides order of magnitude estimates for the cost of production and treating of natural gas.

Gas recovery costs at the wellhead

The economics of gas production is a major subject well beyond the scope of this book. There are many texts available, as well as articles published by the Society of Petroleum Engineers.[32] The objective of this crude analysis is to give order of magnitude estimates for the production cost of the raw (wellhead) gas.

Well costs vary markedly with location, the nature of the reservoir, and the type of development required. Development costs in U.S. dollars range from about $1 million/10 Bcf ($1 million/10 J) to $15 million/10 Bcf ($15 million/10 PJ). An onshore well may typically cost $1–$2 million, with offshore wells costing from $3–$20 million, depending upon such factors as water depth, etc. For fields in very difficult terrain, such as Papua New Guinea, the difficulties encountered can reverse these figures. A well in the inhospitable jungles and mountainous terrain may cost $10–$20 million, while wells in the shallow waters of the coastal basin cost typically $3.0–$8.0 million.

For example, Petzet reports statistics for gas production at Pinedale in the Green River Basin of southwestern Wyoming (table 3–5).[33]

Table 3–5. Typical well-head costs in central USA

Well Cost	$MM 4.1
Reserves per well	10Bcf
Finding and development cost	$0.5–0.6/Mcf
Reserve life	40 years

Source: A. Petzet. "U.S. E&D Operations Focus More on Gas, Resource Plays," *Oil and Gas Journal*, April 19, 2004, p. 36; and the Ultra Petroleum Web site: www.ultrapetroleum.com

Well productivity depends on many factors, including depth. In deeper wells with higher pressures, more gas is produced. Gas wells usually have a life cycle comprising three phases:

- An initial production phase that represents the buildup of production to a plateau period.
- The plateau period of typically five years, during which the well is managed so as to produce the optimum output.
- A period of decline, leading to the point at which the well is abandoned. During this phase, the well may be worked over and its productivity restored, thus extending the life of the well.

For a given project there will be a series of wells, with some coming into production, a number in the plateau phase, and some in decline. The objective of the operator is to maintain constant output by judicious management of the well cycles.

For example, one could consider a project that will require the production of about 40 PJ/y from an onshore development where the gas well typically produces 5 MMscfd of gas of 1,000 BTU for five years. If the cost of each well is US$2 million, the project would have the statistics given in table 3–6.

Table 3–6. Statistics for hypothetical on-shore development

Well production rate	MMscfd	5
Gas HHV	BTU/cf	1000
	PJ/a	1.93
Number of wells in production		21
Annual production	PJ/a	40.4
Well cost (each)	$MM	2
No. of wells required over 20y		84
Total cost	$MM	168
Capital Recovery (15% Capex)	$MM/a	25.2
Operating costs (10% Capex)	$MM/a	16.8
Total costs	$MM/a	42
Gas production cost	$/GJ	1.04
	$/MMBTU	1.01

Now consider a project producing gas from a large offshore reservoir. Analysis of the published data from the Australian North West Shelf project indicates that at peak production, the North Rankin gas platform produced 333 PJ/y from 34 production wells.[34] This platform cost US$490 million (Aus$654 million). For a long-term (20-year) development, two such platforms would be required. This type of project gives the statistics shown in table 3–7.

Table 3–7. Statistics for a hypothetical offshore facility

Well production rate	PJ/a	9.79
Number of wells in production		34
Annual production	PJ/a	333
Total cost	$MM	900
Capital Recovery (15% Capex)	$MM/a	135
Operating costs (10% Capex)	$MM/a	90
Total Costs	$MM/a	225
Gas production cost	$/GJ	0.68
	$/MMBTU	0.71

What this order of magnitude estimate shows is that gas from both small onshore fields and large offshore fields can be produced at the wellhead for about $0.50–$1.00/GJ (MMBTU).

However, these costs take no account of the development costs of a gas field. The development costs would include exploration costs, possibly going back several decades. Gas exploration companies and producers capitalize these costs into the development costs and obviously want to recover these costs at an adequate rate of return. These development costs can more than double the production costs derived simply above.

Wellhead gas costs. In other words, the cost in U.S. dollars of buying gas at the wellhead would be typically $1.20–$2.00/GJ (MMBTU), to which is added the cost of treatment. This treatment cost would range around $0.10–$1.00/GJ (MMBTU), depending on the complexity of the treatment required.

The U.S. Energy Information Administration (EIA) collates data for the wellhead gas price. The data are shown in figure 3–10.

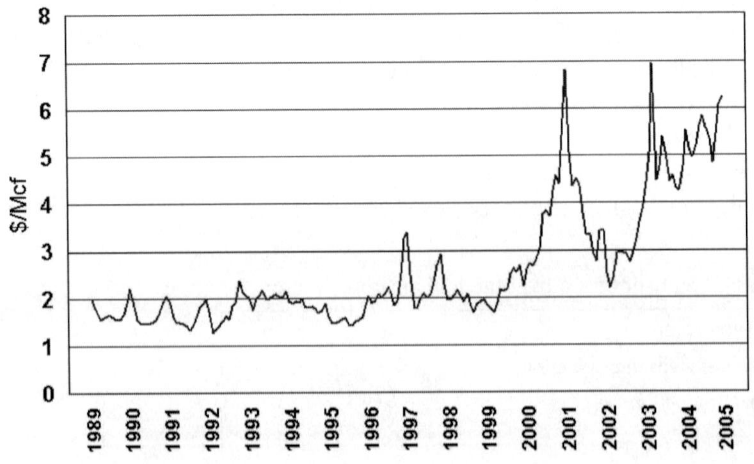

Fig. 3–10. U.S. wellhead gas prices ($/Mcf). Source: U.S. EIA.

This graph shows collated data from across the United States and includes the value of NGL from 1989 to 2004. Prior to about 2000, apart from a few spikes, the prices ranged typically from $1/Mcf to $2/Mcf. At about 1,000 BTU/cf, this price range is in agreement with the above cost analysis. Since 2000, gas prices have been very volatile. As with the rise in energy prices over the period 2003 onwards, there has been a dramatic rise in the price of wellhead gas.

The EIA has further analyzed its data and critically compared the data with the spot prices at the Henry Hub.[35] Henry Hub prices are for gas stripped of NGL and delivered to the Henry Hub Gas Processing Plant. Over the period 1996 to 2000, they found a strong correlation (correlation coefficient 0.975), with the Henry Hub spot price being about 10% higher than the wellhead price. This difference represents the cost of transport to the center.

Case study: economics of large gas plants

The economics of large gas plants (> 1,000 MMscfd gas) is of importance in an understanding of the economic drivers in the refinery and petrochemicals operations. Because of the large flow of gas, these plants produce large volumes of natural gas liquids.

Natural gas condensate from these operations, often called natural gasoline, can be used directly as blendstock for gasoline production. Its value to the gas plant operation is intimately linked to the prevailing price of crude oil via the value of gasoline. LPG (propane and butane) is also linked to the prevailing price of crude by the energy market. There are some seasonal factors, so the linkage is not as direct.

Ethane (and also propane and butane) is used as a feedstock for the production of ethylene. For this role it competes with naphtha, which has a direct relationship with oil price.

Large gas plants often have the advantage that when the price of naphtha (oil) is low relative to the price of gas, ethane can be left in the gas stream and sold at the gas price, thus saving the extraction cost. Conversely, in time of low gas price and high oil price, ethane can be extracted and profitably sold.

This next case study will consider a large gas plant with data as listed in table 3–8.

Table 3–8. Statistics for a large hypothetical gas-plant

		Input Gas	Sales Gas
Flow	MMscfd	1000	
Flow	PJ/a	450	286
Methane	vol%	80.20%	94.90%
Ethane	vol%	7.00%	1.66%
Propane	vol%	4.40%	0.26%
Butane	vol%	2.30%	0.03%
C5+	vol%	3.40%	0%
Inerts	vol%	2.70%	3.19%
Liquids	**(t/a)**	**PJ/a**	
Ethane	650103.9	33.72	
Propane	711569.01	35.83	
Butane	510951.7	25.3	
C5+	947071.01	46.42	

Analysis of recent published data for the construction cost of large greenfield gas plants indicates a cost of $950 million.[36] The plant would have economic parameters as given in table 3–9.

Table 3–9. Economic statistics for a large gas plant

CAPEX		$MM	950
Capital recovery (10% DCF, 20y, FACTOR 14.3%)		$MM/a	135.85
OPEX (5% CAPEX)		$MM/a	47.5
Well head gas input		MMcf/d	1000
Wellhead gas cost		$/GJ	2
INPUTS	**MMt/a**	**PJ/a**	**$MM/a**
Process gas	8.409	427.52	855.03
Fuel and losses (5%)	0.443	22.50	45
Total inputs	8.851	450.02	900.03
Outputs			
Ethane	0.650	33.72	
Propane	0.712	35.83	
Butane	0.511	25.30	
Sales gas	5.589	286.25	
Gasoline (C5+)	0.947	46.42	
Total outputs	8.410	427.52	
Annual costs			1083.38
Production cost ($/GJ)			2.53
		$/t	**$/bbl**
Ethane		131.52	7.45
Propane		127.47	10.28
Butane		124.17	11.53
Gasoline		117.58	13.74

This illustrates that for an input wellhead gas price of $2.00/GJ, the production cost of the sales gas is $2.50/GJ. Of interest are the concomitant production costs of LPG and gasoline, which are well below the prevailing prices of products derived from crude oil. This makes such operations extremely profitable at high oil prices. Plants with these statistics occur in the Middle East and other areas where there is no tangible link between the wellhead price and the prevailing price of energy (oil).

At a wellhead price of $6.00/GJ, the production cost of the sales gas and the products are $6.74/GJ. This is more typical of the case in the United States and Europe, where the prices of wellhead gas are linked to prevailing energy prices. This lowers the operating margin of the plant.

Greenhouse Emissions

The Kyoto Protocol aims to limit global warming by capping and reducing the amount of greenhouse gases emitted to the atmosphere. The principal greenhouse gas is carbon dioxide, which is used as the standard. Methane, which has 21 times the greenhouse impact of carbon dioxide, and nitrous oxide, with 310 times the effect of carbon dioxide, are also included in the Protocol. Few other carbon emissions are considered from the Kyoto standpoint, even though these may ultimately lead to carbon dioxide in the atmosphere (such as carbon monoxide or ethane). However, such gases are often controlled by other environmental regulations.

As part of the process of greenhouse gas (GHG) emission reduction, regulatory authorities have a variety of legislative instruments. These range from prohibiting emission activities to forcing developments to use low-emitting technologies. They also include applying taxes to emitted carbon dioxide and pressuring companies to enter a carbon trading scheme.

The earlier years of the debate were characterized by many gas companies embracing the Kyoto agenda. This was done on the assumption that it would benefit gas over its competitor, oil and oil derivatives, and in particular, over coal. This arises because upon combustion, gas (methane) has a much lower impact than alternative fossil fuels. This is illustrated in table 3–10, which gives pertinent emission factors.

Table 3–10. Greenhouse gas emission for various fuels

Fuel	% Carbon	Tonne Carbon/TJ	Tonne CO_2/TJ
Natural Gas	76	14	51.3
LPG	81	16.4	59.4
Naphtha	87	18.2	66
Fuel Oil	89	19.2	69.7
Brown Coal	25	26.2	95
Black Coal	67	24.8	90
Wood	42	25.9	94
Bagasse	26	26.7	96.8

As can be seen, natural gas produces less carbon dioxide emission than LPG, naphtha, oil, and coal. Thus moving from coal- to gas-fired power generation (as was widely done in the United Kingdom after 1990) produces a net reduction of GHG emissions. Although wood and bagasse (sugar cane residue) produce higher emissions, these are considered as renewable fuels with zero or near-zero emissions.

A problem with this simple approach is that it ignores emissions at the gas plant. As can be seen from the data on the composition of the wellhead gas, carbon dioxide can be present in gases in very large amounts. Indeed, substitution of coal by gas may result in no net GHG emission saving because of these emissions. This is one aspect of balance sheet shifting that currently plagues the approach to Kyoto at all levels.

The problem of large emissions from gas plants has had the effect of delaying a large-scale gas development in Indonesia (Natuna). This field contains more than 200 Tcf of gas but is 70% carbon dioxide. To date, no economically viable method has been found of developing the field without commensurate GHG emissions.

On a smaller scale, carbon dioxide is separated from wellhead gas on the Snorhit platform in the North Sea. This is then reinjected into a saline aquifer below the gas-producing wells, thereby eliminating carbon dioxide emissions from a shore-based gas plant.

Although gas plants emit carbon dioxide, the gas is concentrated in the effluent from the solvent regenerator streams. This makes the capture of the gas low cost and, if available, the gas can be reinjected into a suitable underground reservoir.

This is well-known in EOR programs, where carbon dioxide is captured from a gas plant or from a gas well with high carbon dioxide content. It is compressed to a liquid and injected into oil-bearing strata to improve oil recovery.[37] This is practiced in many parts of the world.

This concept is being followed in the proposed Gorgon LNG project in Australia. The Gorgon field, containing 10 Tcf of gas, lies off the northwest coast of Australia and is slated for a major LNG development. The wellhead gas contains about 10% carbon dioxide.

The separation and emission of this amount of carbon dioxide to the atmosphere would be more than 1 million tonnes (MMt) for a world-scale LNG operation.

The project aims to pass the wellhead gas to a gas plant, which would separate, then compress to a liquid, and finally reinject the carbon dioxide into an underground reservoir.

Although the scheme has gained government approval, it has other environmental consequences. The best site for the operation, a world-scale LNG plant, is on the environmentally sensitive Barrow Island, where suitable reservoirs for the separated carbon dioxide are found. Old oil and gas operations on the island have resulted in the underlying strata being well characterized by the industry. Alternative sites on the mainland do not have suitable reservoirs.

Both the Snorhit and Gorgon developments are versions of carbon geosequestration. At present this is seen as the best method for stopping carbon dioxide emissions to the atmosphere.

Carbon geosequestration

The concept of carbon geosequestration is to capture carbon dioxide in a gas plant activity, compress the gas to a liquid, and inject the liquid into an underground reservoir. The reservoir would have the characteristics that it is at a depth and pressure to maintain the carbon dioxide as liquid. It would be sealed to prevent long-term migration of carbon dioxide to the surface.

Depleted oil and gas reserves are obvious candidates for these sites, but it is harder to find other suitable sites because of the difficulty in ensuring the reservoir will not leak to the surface.

One of the problems with developing geosequestration is the very high cost of demonstration (proof of concept) projects. Table 3–11 gives some options being developed for geosequestration in the United Kingdom in association with an EOR operation.[38] The association with an EOR project serves to lower the unit cost (in U.S. dollars) of carbon dioxide sequestration from \$50–\$63/t to \$8–\$20/t carbon dioxide.

Table 3–11. Approach to geosequestration in North Sea

Option	Benefits	Problems
EOR applied to an on-shore field	Test bed for developing EOR monitoring and verification techniques	CAPEX US$20–40MM
	Provides experience with regulation and monitoring	
	Provides test-site for long term CO_2 storage	
	Increases public awareness and increases confidence in the technology	
Naturally produced CO_2 EOR in North Sea	Full-scale project 1–2 MMt/a CO_2/a	CAPEX > US$500MM
	Test bed for developing EOR monitoring and verification techniques	Does not demonstrate CO_2 capture technology on a combustion plant.
	Provides experience with regulation and monitoring	
	Provides test-site for long term CO_2 storage	
	Increases public awareness and increases confidence in the technology	
Captured CO_2 EOR in the North Sea	Full-scale project 1–2 MMt/a CO_2/a	CAPEX > US$700MM
	Test bed for developing EOR monitoring and verification techniques	Does not demonstrate CO_2 capture technology on a combustion plant.
	Provides experience with regulation and monitoring	
	Provides test-site for long term CO_2 storage	
	Increases public awareness and increases confidence in the technology	
Captured CO_2 from a power plant for EOR in the North Sea	Full-scale project 1–2 MMt/a CO_2/a	CAPEX > US$1,100MM
	Includes a demonstration of full scale capture	
	Test bed for developing EOR monitoring and verification techniques	
	Provides experience with regulation and monitoring	
	Provides test-site for long term CO_2 storage	
	Increases public awareness and increases confidence in the technology	

Capture costs of carbon dioxide

Amine type plants can readily reduce the carbon dioxide in natural gas to below 2%. The principal problem for sequestration is the extraction of carbon dioxide from flue gases that come from the power generating or other fired-furnace operations. The basic problem is illustrated in figure 3–11.

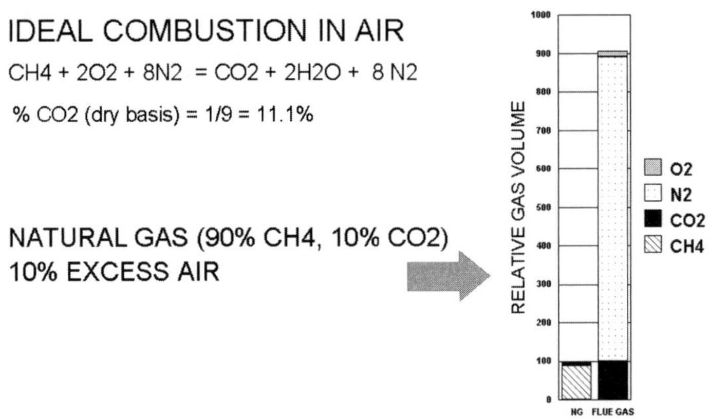

Fig. 3–11. Carbon dioxide content of flue gas

This figure illustrates the volume expansion that occurs on combustion and the volumes of gas that would have to be processed in a normal flue gas capture regime. Typically the carbon dioxide content ranges from 3% to about 15% of the total gas volume. In addition, flue gases contain NO_x and sulfur compounds that have to be removed before the absorber.

The preferred absorbent for this type of operation is monoethanolamine. This sorbent can be degraded by the excess oxygen. The absorbent plants for flue gas duty will require an absorbent regenerator system.

Capture costs from variously sourced fuel gases have been addressed by Friedman and others who have compared some of the alternative proposals for extracting carbon dioxide from flue gas.[39] The aim of the work was to identify the cost of producing carbon dioxide for EOR operations. These indicate a cost in the range of $17/t to more than $50/t for the cost of carbon dioxide capture from flue gases. Table 3–12 summarizes the results.

Table 3–12. Capture of carbon dioxide from flue gas (after Friedman)

Unit Type	CO_2 in Flue	CO_2 MMscfd	CAPEX $MM	ROC $/Mscf CO_2	OPEX $/Mscf CO_2	Total $/Mscf CO_2	Total $/t
Gas Turbine	3.50%	19	50.4	1.16	1.6	2.76	49.72
Coal Fired Boiler	13.0%	19	34.4	0.79	1.26	2.05	36.93
Gas Turbines (2 x 90MW)	3.50%	50.9	130	1.12	0.5	1.62	29.18
Multiple Sources		108.8	476	1.92	0.98	2.9	52.24
Gas Fired Boiler		120	180	0.66	0.31	0.97	17.47

Adapted from B. M. Friedman, R. J. Wissbaum, and S. P. Anderson. "Various Recovery Processes Supply CO_2 for EOR Projects," *Oil & Gas Journal*, August 23, 2004, p. 37.

Present technology cannot economically scrub the carbon dioxide from the flue gases. There are two approaches being developed. Research is being conducted on increasing the strength of the absorbents, which is feasible. However, this route would require increased energy to regenerate the absorbent.

Another approach is to consider the conversion of fossil fuel plants to oxygen as opposed to air-based systems. Thus air-based furnaces would be replaced with oxygen-fired boilers. This produces a flue gas substantially free of nitrogen, and the carbon dioxide could then be more economically captured.

Sequestration costs

Some idea of the cost of sequestration costs can be developed from data available from EOR operations.[40] Salient data are developed in table 3–13 for two scales of operation: 233 kt/y carbon dioxide, typical for EOR projects, and a scaled-up option for sequestration from a large processing facility.

The data indicate that sequestration of captured carbon dioxide will cost (in U.S. dollars) between $20/t and $30/t of carbon dioxide.

Table 3–13. Statistics for CO_2 sequestration

CO_2 injected	kt/a	233.16	1000
Capital Cost	$MM	14.98	38.6
ROC (2y construction, 10% Dcf)	15.20%	2.30	5.97
Operating Costs	$MM/a	1.12	2.89
CO_2 extraction cost ($0.75/Mcf)	$MM/a	3.15	13.51
Total Costs	$MM/a	6.57	22.38
Unit sequestration cost	$/t	28.2	22.38

Methane emissions

Also of concern to the gas industry is the emission of methane. Methane as a GHG has an impact 21 times that of carbon dioxide. Methane emissions arise from fugitive emissions and operations in the gas development and delivery chain. Good housekeeping practices and modest capital expenditures can result in significant reduction in emissions with concomitant increase in company profits.[41]

Of interest are the country-by-country estimates for methane emission by the U.S. EPA and the U.S. EIA. These estimates are shown in figure 3–12 in terms of millions of tonnes of carbon dioxide equivalent ($MMtCO_2e$).

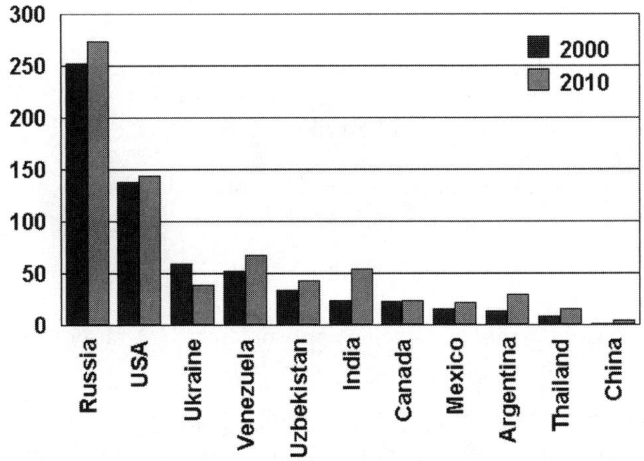

Fig. 3–12. Methane emissions from oil and gas infrastructure ($MMtCO_2e$). Source: U.S. EPA and U.S. EIA.

This illustrates that some nations have very large emissions of methane. Much of this is easily lowered. For example, in many parts of the world, methane from oil and gas operations is vented to atmosphere. Burning the gas by a flare or thermal oxidizer would convert the methane to carbon dioxide, greatly reducing the greenhouse impact.

References

1. Newman, S. A., editor. *Acid and Sour Gas Treating Processes*. Houston, TX: Gulf Publishing, 1995.

2. "Gas Processes" is published biennially in even years in the April/May editions of *Hydrocarbon Processing*.

3. The detailed behavior of the formation of hydrates with natural gas and seawater, and the impact of addition of methanol, is described in papers by Y. E. Makogon and S. A. Holditch. "Lab Work Clarifies Gas Hydrate Formation, Dissociation," *Oil & Gas Journal*, February 5, 2001, p. 47; and ibid. "Experiments Illustrate Hydrate Morphology, Kinetics," February 12, 2001, p. 45. Additional information is given by J. E. Paez, et al. "Problems in Gas Hydrates: Practical Guidelines for Field Remediation," Society of Petroleum Engineers, Technical Paper, SPE 69424; different hydrate inhibitors are compared by J. Davalath and J. W. Baker. "Hydrate Inhibition Design for Deepwater Completions," *SPE Drilling and Completion*, June 1995, p. 115.

4. Methanol can also be used as is described by P. Hampton, et al. "Liquid–Liquid Separation Technology Improves IFPEXOL Process Economics," *Oil & Gas Journal*, April 16, 2001, p. 54.

5. Kirchgessner, D. A., et al. "Advanced Dehydrator Design Recovers Gas, Reduces Emissions," *Oil & Gas Journal*, July 26, 2004, p. 52.

6. Healy, M. J., A. J. Finn, and L. Halford. "U.K. Nitrogen-Removal Plant Starts up," *Oil & Gas Journal*, February 1, 1999, p. 36. They describe a 200-MMscfd plant for reducing approximately 10% nitrogen to < 5% nitrogen at a cost of £40 million (US$80 million).

7. Butts, R. C., K. Chou, and B. Slaton. "Nitrogen-Rejection Process Developed for Small Fields," *Oil & Gas Journal*, March 13, 1995, p. 92. Costs are claimed to be in the range of $0.30 to $0.45/Mcf.

8. Advanced Extraction Technologies, Mehra Process NRU, Gas Processes 2000, *Hydrocarbon Processing*, April 2000, p. 78; Y. R. Mehra, G. C. Wood, and M. M. Ross. "Noncryogenic N_2-Rejection Process Gets Hugoton Field Test," *Oil & Gas Journal*, May 24, 1993, p. 62.

9. UOP NITREX Process, Gas Processes 98, *Hydrocarbon Processing,* April 1998, p. 116; and Englehard Corp., Molecular Gate, Gas Processes 2002, *Hydrocarbon Processing*, May 2002, p. 72.

10. Howard, M., and A. Sargent. "Texas Gas Plant Faces Ongoing Battle with Oxygen Contamination," *Oil & Gas Journal*, July 23, 2001; Society of Petroleum Engineers, Specialist Discussion on the Internet, March 2003.

11. "Unocal Solves Mercury Problem off Thailand," *Oil & Gas Journal*, November 16, 1998, p. 34.

12. Calgon Carbon Company. Gas Processes 94, *Hydrocarbon Processing*, April 1994, p. 92; Costs are given at $300 to $1,200 per MMscfd; UOP, costs are given as $600 to $1000/Mcf. Operating experience is described by D. L. Lund. "Wyoming Operator Solves Mercury Exposure Problems," *Oil & Gas Journal*, May 13, 1996, p. 70, and E. F. Rhodes, P.J. Openshaw, and P. J. H. Carnell. "Fixed-Bed Technology Purifies Rich Gas with H_2S, Hg," *Oil & Gas Journal*, May 31, 1999, p. 58.

13. Axens' Multibed and Syntec's Puraspec technologies in "Gas Processes 2002," *Hydrocarbon Processing,* May 2002, pp. 71 and 74.

14. For example see Shell Global Solutions—Sulphinol technology, "Gas Processes 2002," *Hydrocarbon Processing,* May 2002, p. 78. The operation of an amine absorbent process in combination with the UOP-Selectox process is described by S. G. Jones and R. V. Bertram. "Long-term Operating Data Shed Light on Selectox Process," *Oil & Gas Journal*, August 27, 2001. p. 44.

15. Watson, S., R. Kimmitt, and R. B. Rhenesmith. "Study Compares COS-Removal Processes," *Oil & Gas Journal*, September 22, 2003, p. 66.

16. Newman, editor. *Acid and Sour Gas Treating Processes.* 1995.

17. There are several proprietary variants. See M. V. Twigg, editor. *Catalyst Handbook*, second edition, Wolfe Publishing, 1989, for a detailed description of zinc oxide absorbents. The Iron Sponge Process is described in "Gas Processes 2002," *Hydrocarbon Processing*, May 2002, p. 72.

18. Nilsen, F. P., I. S. L. Nilsen, and H. Lidal. "Novel Contacting Technology Selectively Removes H_2S," *Oil & Gas Journal*, May 13, 2002, p. 56.

19. Palla, N., et al. "Morphysorb Process Proves Feasible in First Commercial Plant," *Oil & Gas Journal*, July 5, 2004, p. 54.

20. Jones, S. G., D. R. Rosa, and J. E. Johnson. "Lisbon Gas Plant Installs Acid-Gas Enrichment, Injection Facility," *Oil & Gas Journal*, March 1, 2004, p. 54, and ibid., "Acid-Gas Injection Design Requires Numerous Considerations," March 8, 2004, p. 45.

21. Shaw, T. P., and P. W. Hughes. "Optimize CO_2 Removal," *Hydrocarbon Processing*, May 2001, p. 53; and A. Habibullah. "Alaska North Slope LNG Project Considers Various CO_2 Removal Processes," *Oil & Gas Journal*, June 3, 2002, p. 46.

22. Mak, J., D. et al. "Consider Physical Solvents to Treat Natural Gas," *Hydrocarbon Processing*, June 2003, p. 87.

23. Toshima, N., editor. *Polymers for Gas Separation*, New York: VCH Publishers, 1992.

24. Lee, A. L., et al. "Membrane Process for CO_2 Removal Tested at Texas Plant," *Oil & Gas Journal*, January 31, 1994. The authors give technical detail for the operation of a membrane separator at different loadings of carbon dioxide and claim costs of $0.10 to $0.25/Mcf.

25. Wills, J., M. Shemaria, and M. J. Mitariten. "Pipeline-Quality Natural Gas after Molecular-Gate CO_2 Removal," SPE Technical Paper SPE 80602, abridged in *Journal of Petroleum Technology,* September 2003, p. 77; Englehard Corporation—Molecular Gate, Gas Processes 2002, *Hydrocarbon Processing,* May 2002, p. 71.

26. Aggarwal, V., and S. Singh. "Improve NGL Recovery," *Hydrocarbon Processing*, May 2001, p. 41.

27. Chebbi, R., et al. "Simulation Study Compares Ethane Recovery in Turboexpander Processes," *Oil & Gas Journal*, January 26, 2004, p. 64.

28. Rahman, A. A., et al. "Petronas Improves Ethane Extraction of Gas Processing Complex," *Oil & Gas Journal*, October 25, 2004, p. 58.

29. Hawkins, D. J. "Alberta Gas Processing Shows Significant Development," *Oil & Gas Journal*, December 16, 2002, p. 46; ibid. "Alberta Straddle Plants Could Process Alaskan Gas," December 23, 2002, p. 54; ibid. "Aux Sable Plant Makes Significant Impact on Alberta NGL Supply," January 6, 2003, p. 48; and ibid. "Australia Straddle Plant's Operating Strategy Mirrors Aux Sable," January 20, 2003, p. 50.

30. Bothamley, M. "Offshore Processing Options Vary Widely," *Oil & Gas Journal*, December 6, 2004, p. 47.

31. Bramoulle, Y., P. Morin, and J.-Y. Capelle. "Differing Market Quality Specs Challenge LNG Producers," *Oil & Gas Journal*, October 11, 2004, p. 48.

32. For example, P. Newendorp and J. Schuyler. *Decision Analysis for Petroleum Exploration*. Aurora, CO: Planning Press, 2000.

33. Petzet, A. "U.S. E&D Operations Focus More on Gas, Resource Plays," *Oil & Gas Journal*, April 19, 2004, p. 36; and Ultra Petroleum Web site: www.ultrapetroleum.com

34. Woodside Limited, Annual Reports, 1990 to 1995.

35. Pudzik, P. "U.S. Natural Gas Markets: Relationship between Henry Hub Spot Prices and U.S. Wellhead Prices," U.S. Energy Information Administration, August 2002.

36. Author's analysis of *Hydrocarbon Processing,* "Boxscore" supplements published February and October 2000 to 2004.

37. Friedman, B. M., R. J. Wissbaum, and S. P. Anderson. "Various Recovery Processes Supply CO_2 for EOR Projects," *Oil & Gas Journal*, August 23, 2004, p. 37.

38. Adapted from report of UK DTI options for geosequestration demonstration in North Sea; "DTI Details Plans for Implementing EOR in North Sea," *Oil & Gas Journal,* May 17, 2004, p. 48; US$2 = 1GBP

39. Friedman, Wissbaum, and Anderson. *Oil & Gas Journal*, August 23, 2004, p. 37.

40. Moritis, G. "CO_2 Sequestration Adds New Dimension to Oil, Gas Production," *Oil & Gas Journal*, March 3, 2003, p. 39.

41. Fernandez, R., D. Lieberman, and D. Robinson. "U.S. Natural Gas STAR Program Success Points to Global Opportunities to Cut Methane Emissions Cost-Effectively," *Oil & Gas Journal,* July 12, 2004, p. 18.

4

Comparison of Energy Fuels

This chapter is concerned with the position of gas as an energy fuel in comparison to alternative fuels.

In energy markets there is usually more than one fuel type available to a user. The fuel choice generally comes down to a question of price. This is very dependent upon the local market and specific location factors. This chapter reviews the competitive position of gas in relation to other energy fuels. It also gives some indication as to when gas will displace other fossil fuels and energy sources and when gas usage may be under threat from other fuel sources.

Natural Gas Use: Project Scale of Operation

The intended use of gas and its competitive position versus other fuels is dependent upon the scale of operation.

If a gas development project is small (10–20 MMscfd; 3–7 PJ/y), the likely target market may be a small town (reticulated gas for cooking and heating). It could also be a power plant for local power (around 50 MW) or industry. Local power generation is attractive when supplying a peak demand is an issue. Some moderately energy intensive industries (glass, foundries, etc., including cogeneration projects) can consume 3–5 PJ/y of gas. As well as competition from conventional petroleum fuels (naphtha, diesel, and fuel oil), natural gas in these markets is often in competition with alternative fuels, such as waste oil and biofuels.

If a gas development project is of a moderate scale (100–200 MMscfd; 30–70 PJ/y), the likely target markets are town gas (reticulated gas to domestic, commercial, and industrial users) or power generation (about 500 MW). This would include both baseload and peak power projects. This volume of gas could also be used to supply a major energy-intensive enterprise (incorporating cogeneration, such as alumina refining). It could also be used to manufacture methanol or ammonia in a world-scale plant. For the most part, the gas in such projects is in competition with the liquid petroleum fuels.

For very large scale developments (> 500 MMscfd; > 150 PJ/y), operations are necessarily very large. These include providing reticulated gas to major cities and urban centers, several world-scale chemical plant operations, or a large export LNG project. For these projects, the major competitor with gas is coal, either as coal itself or coal-derived electricity or synthetic natural gas (SNG).

Properties of alternative fuels: coal and petroleum products

Properties of liquid fuels. Table 4–1 lists the common conversion factors for petroleum liquids. Each of the liquid classes exists over a range of densities and physical properties. The conversion factors are given for a typical product of international trade. Reporting agencies (such as Platts, Reuters, etc.) use factors such as these to correct prices to a common basis. The energy content is for the HHV, also called the gross heating value. As a measure, a tank of gasoline (60 L) contains approximately 2 GJ of fuel.

Table 4–1. Common conversion factors for petroleum fuels

	kg/litre	bbl/tonne	HHV (GJ/tonne)
LPG	0.519	12.10	50.0
Naphtha	0.700	9.00	49.6
Kerosene	0.793	7.94	46.4
Motor Diesel	0.846	7.44	45.6
Fuel Oil (LSWR)	0.888	7.09	44.0
Fuel Oil (HSFO)	0.973	6.46	42.0
Crude Oils			
43.2 API	0.793	7.94	46.1
40 API	0.825	7.60	46.1
35 API	0.850	7.40	45.0

Properties of coals. The typical properties of various coals and other solid fuels in competition with gas are listed in table 4–2. Further information on coal properties is given in appendix B.

Table 4–2. Typical properties of coal and solid fuels

	HHV (GJ/tonne)
Export coking coal	29–32
Export steaming coal	27
Power generation coal	21–24
Brown coal	9-11
Briquettes	22.1
Wood (dry)	16.2
Bagasse	9.6

By comparison to liquid petroleum fuels, solid fuels have lower energy per unit mass (specific energy content). Further, there is a large variation in values. Black coal ranges from about 21 GJ/t for raw coal, which can be used for power generation, to more than 30 GJ/t for cleaned, export coking coal.

By comparison, brown coal and lignite, which contain a lot of absorbed water, have much lower energy contents (typically 10 GJ/t). The water can be removed by the briquetting process. This increases the energy content to a similar level as that for raw black coal.

Wood has a range of energy contents depending on the type and amount of water absorbed. Dense woods can have an energy content of 16 GJ/t.

Where it is available, bagasse (sugar cane waste) is often used for power generation. Bagasse has energy content of typically 9 to 10 GJ/t.

Heating Values (HHV and LHV)

The energy content of a fuel is most often quoted as the HHV, or gross heating value. This value is determined when the fuel is burnt, and the product water formed in combustion is condensed. If the water of combustion remains in the gaseous state, which is the more common state, then the LHV, also known as net heating value, is the result. Thus the HHV value is higher than the LHV value by the heat liberated when the water of combustion is condensed.

Ratio of heating values

The difference between the HHV and the LHV depends on the hydrogen content of the fuel. Figure 4–1 shows the ratio of HHV to LHV for pure paraffins, which have relatively high hydrogen content.

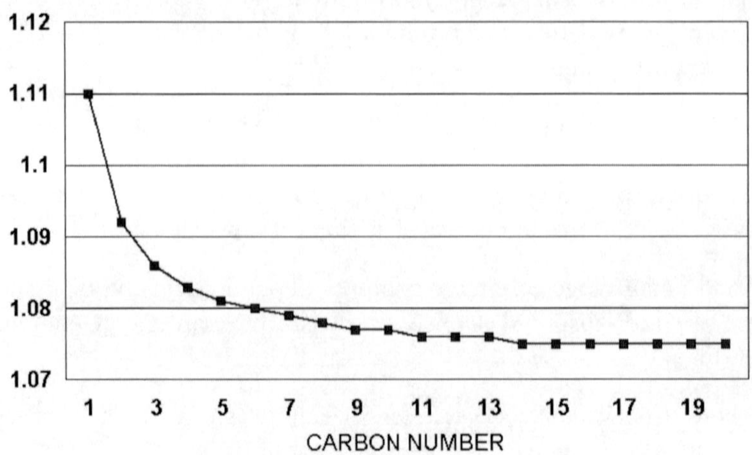

Fig. 4–1. Ratio of HHV to LHV for n-paraffins

The graph shows there is a major difference (more than 10%) between the HHV and LHV for methane (C_1), which is the major component of natural gas. For heavier molecules, the difference is smaller (about 8.5% for LPG, propane, and butane), and it is smaller yet for liquid paraffins (typically about 7.5%).

For liquid fuels, the difference between HHV and LHV is smaller still. This is because of the presence of aromatics, which have a very low hydrogen content. For coal, there is very little difference between the HHV and the LHV values.

Determining liquid fuel heating values

The measurement of heating values is arduous and expensive, and it uses an instrument known as a bomb calorimeter. Heating values are mostly estimated from the chemical composition and physical properties of fuel. For instance, the difference between the HHV and LHV of any fuel is predictable from knowledge of the composition. This could include its carbon, hydrogen, sulfur, inert materials, ash, and water contents. The major factor is the hydrogen content, which for petroleum fuels is related to the density of the fuel.

Figure 4–2 shows the predicted HHV and LHV values for petroleum fuels using the ASTM D-4868 method. The method works well for fuels with densities greater than about 0.8 kg/L(kerosene and heavier fuels). The method is poor for naphtha (gasoline) and lighter fuels.

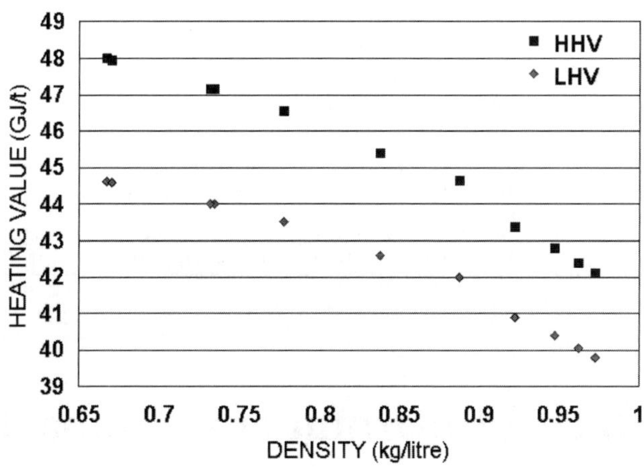

Fig. 4–2. Estimates of HHV and LHV of liquid fuels

It is important to note that prices of gaseous energy fuels are often quoted in terms of HHV alone. However, this may not be the best option for a duty where the water of combustion is not condensed, such as simple cycle gas turbines or transport fuel use. Buying and selling energy in terms of a price related to its HHV builds in an advantage for gas, which may in practice not be realized.

Nonconventional Energy Sources

In many countries there is increasing support for the use of nonconventional energy fuels. In some cases, these fuels compete directly with natural gas, often with a government subsidy. Table 4–3 gives an overview of the competitive advantages and disadvantages of the various nonconventional fuels that compete with natural gas.

Table 4–3. Comparison of some non-conventional fuels

	Advantages	Disadvantages
Brown coal briquettes	Large lignite reserves	Low energy content
Waste oil	Effective disposal of intractable waste	Contaminants—metals, lead etc.
Scrap tires	Low value fuel oil—effective disposal of intractable waste	High cost
Bio gas	Power generation by waste product	High sulphur content—low energy content
Landfill gas	Power generation and disposal of noxious gas	Contaminants—low energy content

Relative Environmental Impact

If present in natural gas, hydrogen sulfide is removed by appropriate gas treatment and is not an issue upon combustion. Sulfurous odorant is added for safety reasons and is typically only about 30 ppm. Nitrogen is not a priority issue, although conventional wisdom suggests that a high nitrogen content fuel results in high nitrogen oxide (NO_x) emissions.

Carbon dioxide content of some natural gases

As discussed in chapter 3, natural gas contains varying quantities of carbon dioxide. To further illustrate the issue, table 4–4 gives representative values for carbon dioxide from various natural gases in Australia. Not only does carbon dioxide vary with each basin, but there is also considerable field-to-field variation.

Table 4–4. Carbon dioxide content of some wellhead gases

Area/Basin	Field	Volume %	Weight %
Gippsland	Barracouta	0.8	1.8
	Marlin	2	3.9
	Kipper	9.5	18.3
Central Australia	Cooper	17	32
	Palm Valley	0.4	2.9
	Mereenie	0.1	1
North West Shelf	Goodwyn	1.4	2.9
	Woodada	4.2	10.3
	Scott Reef	11	23.2

However, note that if the composition is expressed on a weight basis, it is significantly higher in carbon dioxide content. Carbon dioxide is removed in conventional gas treatment plants and then generally emitted to the atmosphere. In discussions concerning the greenhouse effect and mitigation of carbon dioxide emissions, the mass of carbon dioxide emitted is the important parameter. Thus it is necessary to determine the gas source and account for any emissions at the gas plant.

Relative greenhouse emission factors are discussed in chapter 3.

Comparative environmental effects

Table 4–5 gives an overview of the environmental impacts of competitive fuels from the standpoint of NO_x emissions, sulfur emissions, and unburnt hydrocarbons. The beneficial effect of natural gas is evident. However, one should note the relatively high NO_x emissions. These are due principally to the high flame temperature of natural gas combustion, which assists the direct formation of nitrogen oxides from the combustion air.

Table 4–5. Comparative environmental impact of competitive fuels

Fuel	NOx (% N in fuel)	S (% S in fuel)	Hydrocarbons emissions	Other emissions
Natural gas	high (1)	low (0.0030)	low	
Landfill gas	moderate (3–6)	high (1–2)	low	chlorine
Fuel oil	high (0.1)	mod–high (0.2-4)	moderate	
Waste oil	high (0.12)	high (3.8)	moderate	lead
Brown coal	moderate (0.51)	low–mod (<1–3)	moderate	ash
Black coal	moderate (1.5)	mod–high (0.5–4)	moderate	ash
Wood	low	low (0.05)	high	ash
Garbage		mod (0.52)	moderate	chlorine

Relative Prices and Price Linkages

U.S. energy prices

Representative prices for gas, oil, and coal in the United States on an energy basis are shown in figure 4–3. All prices are as reported by the U.S. EIA: gas prices, at the wellhead; oil, West Texas Intermediate (WTI); and coal, from imports into the United States. The data have been converted into energy units (US$/GJ) using standard conversion factors. It is assumed that the wellhead gas has an HHV of 1,050 BTU/cf, WTI is 45 GJ/t, and coal is 28 GJ/t.

The price of crude oil rose from less than $3/GJ (below $15/bbl) in 1998 to more than $6/GJ during 2000. The price then fell back to about $3/GJ level at the beginning of 2002 before rising again to more than $7/GJ (more than $50/bbl) at the end of 2004.

Wellhead gas followed this trend, though the peaks and troughs lagged the oil market by one or two months. In energy terms, even allowing for a 10% increase in price as a consequence of processing, the gas price was always below that for oil except for two spikes in the winters of 2000/01 and 2002/03. An average differential between WTI and gas is about $1.40/GJ.

Coal, like many minerals, is sold under longer term supply contracts. Thus the response of price to supply/demand balances is much slower than either the crude oil spot market or sophisticated gas markets such as in the United States. The values illustrated are yearly average figures, with the data of 2003 and 2004 being quarterly.

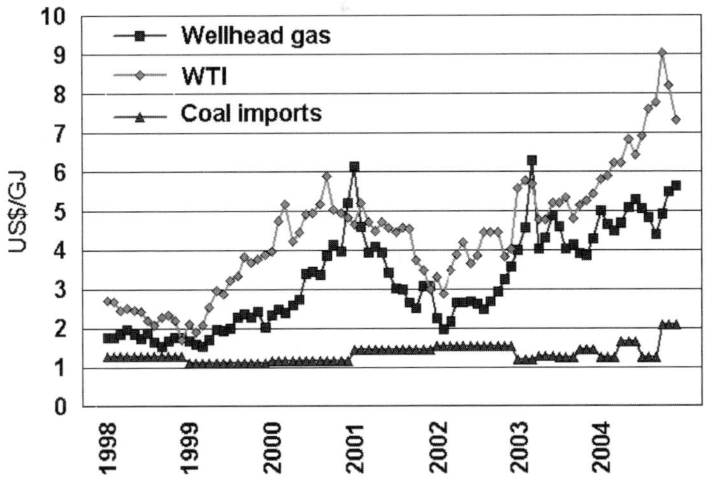

Fig. 4–3. U.S. energy prices

In 1998, at the time of low oil and gas prices, the import coal price was about $1.30/GJ. In comparison, at the end of the period (during a time of high oil prices), coal had only risen to about $2.10/GJ.

Value of the U.S. dollar. The convention is that throughout the world, energy prices are quoted in U.S. dollars. Over the period of interest, there has been a marked deterioration in the value of the U.S. dollar, and no analysis of energy prices would be complete with taking into account this issue. Figure 4–4 replots this data for the U.S. market expressed in Euros from the period since the inception of Euros in 2002.

This graph indicates that although the price of oil has risen, the increase is not as pronounced when expressed in Euros. Further, after the 2002/2003 spike, gas prices have been relatively stable. Coal has shown a decline in value when expressed in Euros.

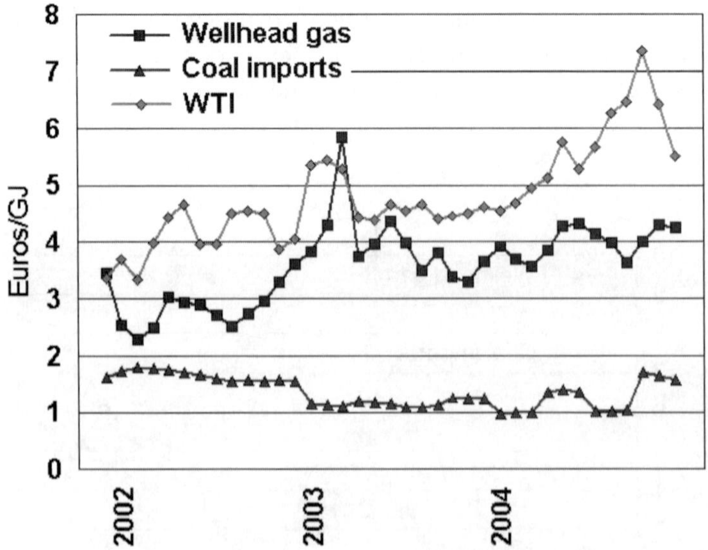

Fig. 4–4. U.S. energy prices in Euros

Energy switching

Although there is a strong linkage between the value of gas and alternative energy fuels, this is not a one-for-one equivalence in energy terms. This is because of the differences in equipment and capital requirement in using an alternative fuel.

In theory there are few fundamental technical obstacles to energy switching between liquid and gaseous fuels. In many parts of the world, power generation plants can switch fuel to take advantage of a low price in any particular feedstock (e.g., gas versus LPG or naphtha). However, this is not the case for most end users. For most users, the decision to fuel switch has long-term implications. Further, it is costly to switch from gas and petroleum fuels to coal, but less costly to convert from coal to gas or petroleum products. Relative to coal and oil, gas has many advantages in terms of transport, storage, and pollution control.

The pricing of natural gas (and its derivatives, LNG, LPG, condensate, and fuel methanol) has to fit into this picture of fuel competition. For power generation, the advantages of natural gas, LNG, and fuel methanol are enhanced over oil derivatives, such as fuel oil. This is due to its easier usage in modern gas turbine plants and its lower sulfur emissions. However, these advantages over oil are not as great as the advantages both have over coal. Therefore coal inevitably competes on price against gas and liquid alternatives.

In addition it should be noted that in many cases, natural gas and LNG are effectively in competition with other fuels. The values of these other energy materials often set the price for gas or LNG.

For instance, the landed price of LNG in the Far East is often related to the price of crude oil via the equivalent price of gas oil (fig. 4–5).

Perhaps the most transparent linkage to prices is for baseload power generation in large (500-MW+) power generating stations. Gas at $2–$3/GJ is competitive with coal at half this value in terms of simple economics. However, gas does have a major advantage in much lower emissions than coal.

Noneconomic factors

There are factors that are not of a commercial nature that help to determine the degree of substitution. Light products—gas, LPG, and naphtha—can be readily substituted for each other in many process operations such as gas turbines. Products in this group would thus be of higher value to a major customer practicing substitution.

Overdependence on imported oil has led many power utilities (and governments) to diversify into LNG and LPG as an energy source for baseload power generation. Such strategic reasons for promoting LNG and LPG give them a tangible benefit over and above the value determined by energy equivalence. Similarly, environmental considerations generate a premium for low-sulfur fuels in relation to coal and the heavier fuel oil grades.

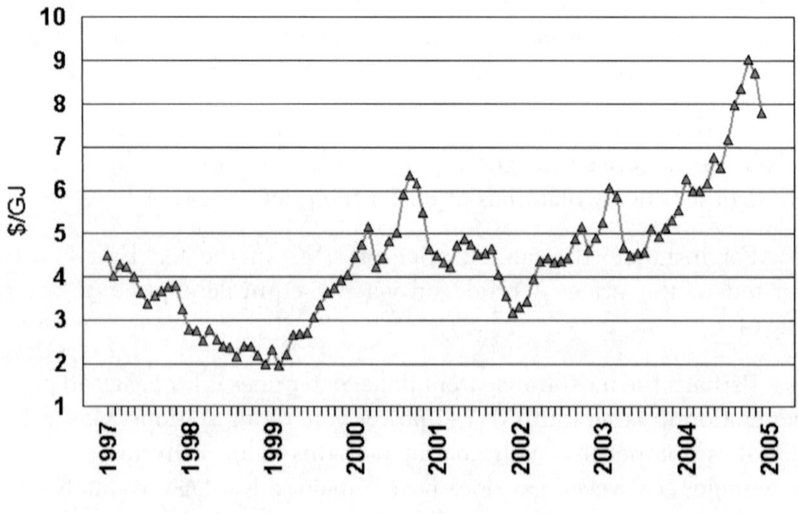

Fig. 4–5. Historical price of gas oil ($/GJ)

5

Gas Transport

There are three modes of mass transport in which the energy (or hydrocarbon content) of natural gas can be transported over long distances to a user:

- By pipeline
- As LNG in specialized ships
- By conversion at or near the wellhead to a liquid product followed by transport by sea tankers—methanol and DME.

This chapter discusses the economics of gas transport through these three methods. An overview of the world's shipping fleet for transporting the products of interest is given.

Comparison of the Shipping Fleets

At present, contract shippers conduct most shipping of liquids (chemical and oil derivatives). The merchant fleet is extensive, and there is a variety of contracts available for the regular movement of product.

Large amounts of liquid product can be moved in large (> 125,000 t) tankers. These are generally referred to as dirty cargoes because the product transported is crude oil and residual fuel oil. Generally this fleet is unsuitable to transport a gas-derived liquid such as methanol, which is a clean cargo. For the transport of these gas products, dedicated ships may be required. This may be provided as a contract arrangement.

By contrast, the clean cargo fleet (chemicals, naphtha, and gasoline) has a wide range of vessel sizes available (10,000 t to more than 100,000 t). Transport fuels are typically moved in loads (parcels) of about 80,000 t, at a cost of typically \$10/t (about \$0.20/GJ). Most chemicals use smaller ships, and costs are higher, typically \$25–\$30/t, or \$1.00–\$1.50/GJ for chemical methanol.

An important point to note is that contract shipping offers financial advantages over owner-operated and dedicated fleets (such as those used to transport LNG). However, the contract price is dependent on the vagaries of the shipping market, which is both cyclic and seasonal.

For LPG, a very large contract merchant fleet is available, although this is dominated by a small number of key players. The available fleet typically moves product in parcels of about 30,000 t to 40,000 t, at a typical cost of \$30–\$40/t, about \$0.60–\$0.80/GJ. However, there are larger ships available (75,000 t).

The cost of contracts is very dependent on business cycles and on the season, as large LPG demand coincides with the northern winter. In order to smooth out the costs (from the ship owner's perspective) most of the fleet is capable of transporting ammonia and other chemicals as well as LPG cargoes. Thus shipping costs also become influenced by the seasonal nature of ammonia (fertilizer) demand, especially the U.S. corn market.

Table 5–1 gives an overview of the merchant fleet available for the transport of energy.

Table 5–1. Comparison of transport fleets for shipping energy

Fleet	LNG	LPG	Chemical	Clean Fuels	Crude oils
Products shipped	LNG only	LPG, ammonia, chemicals	liquid chemicals	naphtha, gasoline, gas oil	crude oils, fuel oil
Size (tonnes)	90,000	10,000–75,000	10,000–40,000	60,000–120,000	> 120,000
Ship types	cryogenic	cryogenic and pressure	sealed tanks	sealed tanks	sealed tanks
Fleet	dedicated	contract	contract	contract and dedicated	contract and dedicated
Cost variation	fixed	seasonal	business cycle	business cycle	business cycle

Solids transport. The transport of solids is conducted in large ocean-going ships and barges with relatively simple off-loading and on-loading machinery. This means that for coal, transoceanic transport costs are relatively low. Typical intercontinental costs are \$10/t (Australia to Japan), or about \$0.33/GJ.

Gas Pipelines

Despite their simplicity, pipelines are highly capital intensive. Laying a pipeline is costly, with the cost split roughly evenly between materials and labor. In addition, provision for compression stations, which are necessary for mass transport over long distances, can contribute 40% of final installed capital costs. Once established, the operations of a pipeline system can cost 5% of the fixed capital per annum.

The capital cost of a pipeline depends upon such factors as pipe diameter, distance, and the amount of compression required. Undersea pipelines cost about double that of land-based pipelines. Operating costs reflect labor charges and fuel usage in compression and compression stations. In some countries, such as Australia and Russia, pipelines can be laid over vast distances at a low cost. This is due to low disturbance and right-of-way (ROW) charges and relatively accessible terrain. In highly urbanized societies, such ROW charges can add considerably to the cost of pipeline construction.

As a rule of thumb, a capital cost (in U.S. dollars) of \$1 million/km can be expected for new land-based pipelines. The capital cost is about \$2 million/km for new undersea pipelines, including compressor stations for countries like Australia. When built, the annual operating cost is typically about 5% of the installed capital.

Laying pipelines in difficult jungle terrain will cost substantially more than this. This is evidenced by the \$1 billion cost for the development of the oil field at Iagafu in Papua New Guinea. A substantial portion of the cost was for the transmission of the product a distance of 274 km to the coast in a 0.5-m pipeline.

The following gives a more detailed analysis of pipeline costs.

Estimates of pipeline cost: uncompressed

The cost of gas transport by pipeline is dependent on the distance travelled and the pipeline diameter. The *Oil & Gas Journal* annually publishes statistics on the construction of pipelines in the United States. Figure 5–1 gives the costs of pipeline per kilometer in the United States during the 10-year period 1995 to 2004 for various pipeline diameters (in inches). The raw data have been extrapolated to 2004 costs using Nelson-Farrar indices.

(For some years, only one pipeline of a specific diameter was constructed. These have been discounted from the data.)

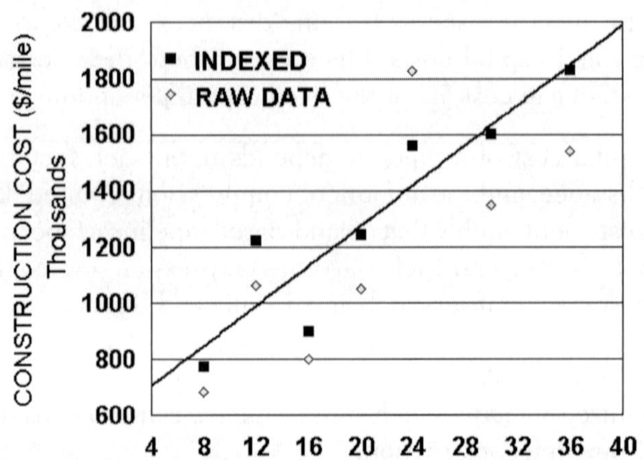

Fig. 5–1. U.S. pipeline costs (corrected to 2004)

The trend line fits a series of the formula:

$$\text{Cost (in million US\$/mile)} = 563{,}000 + 35{,}600 \times D \qquad (5.1)$$

where D is the pipeline diameter in inches.

The equation has a correlation coefficient of 0.848. The equivalent equation in metric units is:

$$\text{Cost (in million US\$/km)} = 350{,}000 + 871{,}000 \times d \qquad (5.2)$$

where d is the pipe diameter in meters.

The cost of pipeline construction has four components: material, labor, ROW and damages costs, and miscellaneous charges. The latter costs include surveying, engineering, supervision, interest, administration overheads, and contingencies. The relative cost components vary considerably depending on the specific pipeline, and even then the average varies from year to year. The breakdown of costs for 2004 in the U.S. market is shown in figure 5–2.

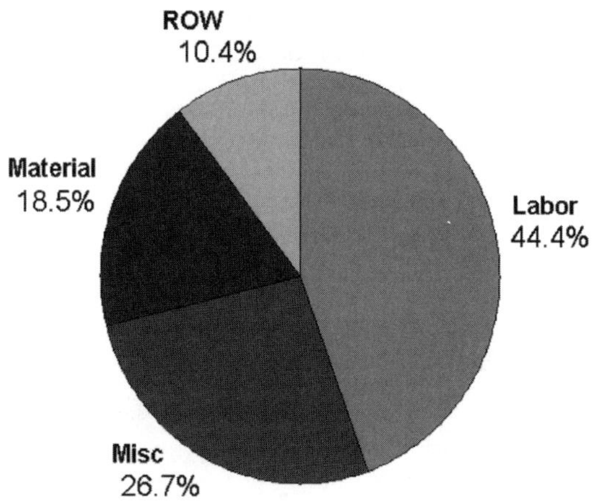

Fig. 5–2. Breakdown of U.S. pipeline costs (2004)

Fig. 5–2 illustrates that the major cost is construction labor, at almost 50% of the total. ROW costs accounted for only about 10%; however, for some pipelines in urban areas, this component cost can be significant. The other costs (about 40%) are made up of material and miscellaneous costs. For 2004, the second largest cost was miscellaneous costs (about 26%), with material costs lower (about 18%).

Pipeline compressor costs

The other major cost for gas pipelines is the cost of compression. Because of the compressibility of gas, pressure drops with distance. In order to overcome pressure drop, recompression stations are placed at intervals in the pipeline, typically at distances of 60 km for a fully compressed line. However, some pipelines, in order to minimize costs, operate without compression, allowing the gas pressure to fall along the pipeline length. This limits the amount of gas that can be transported.

For example, one could consider a 50-km pipeline, with a diameter of 0.61 m (24 in.), carrying 200 PJ of gas at an inlet pressure of 100 bar (10 MPa). In this case, a pressure drop of about 10 bar (1 MPa) will occur. To restore the pressure will require about 1.8 MW. For a fully compressed line, allowance should be made for compressor servicing.

The power required for compression is a function of the throughput. As a guide to the power requirements, figure 5–3 gives the required power for various throughputs, where the duty of the compressor is to raise the pressure by 1 MPa (10 bar). Allowance is made for taking compressors off-line for maintenance.

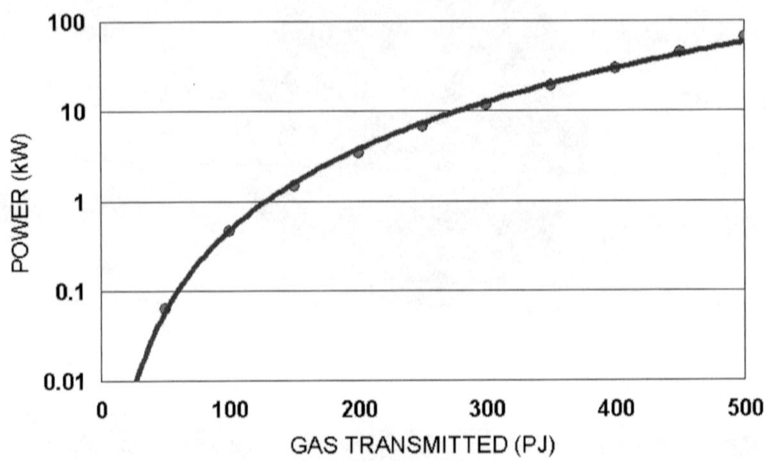

Fig. 5–3. Compressor power and throughput

For ideal operation (100% efficiency), fuel is consumed on an annual basis at the rate of 31.5 GJ/kW (22.3 MMBTU/hp). This fuel is often provided by the gas itself or by diesel. Modern compressor efficiencies are about 80% for multistage compressors operating with interstage cooling to dissipate the compression heat. With recompression every 50 to 60 km or so, long pipelines can consume up to about 3% of the gas transmitted.

Cost of compression. Compression cost rises with power required. However, there is a broad range of installed costs. The data in figure 5–4 are taken from U.S. data in 2003/4 published in the *Oil & Gas Journal*.

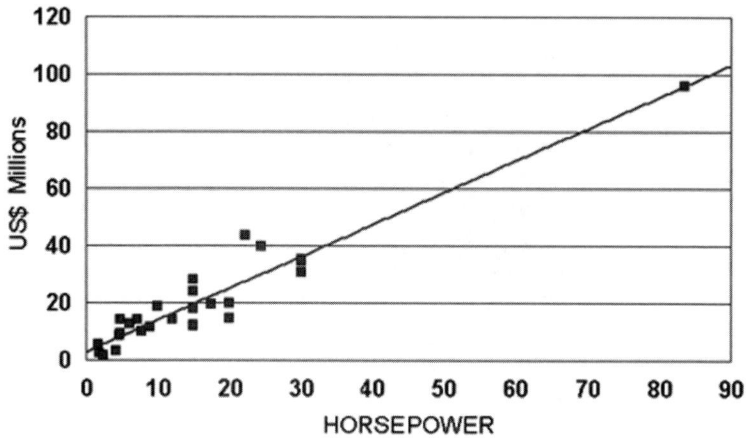

Fig. 5–4. Compressor costs. Source: True, W. R., and J. Stell. "Special Report—Pipeline Economics," *Oil & Gas Journal,* August 23, 2004, p. 52.

There is a considerable scatter in the base data, and the line shown is the result of regression analysis. The line in the figure is for the equation:

$$\text{Compressor cost (in million US\$)} = 2{,}970{,}000 + 1{,}120P \qquad (5.3)$$

where P is the compressor power in horsepower.

This equation has a correlation coefficient for this data set of 0.91. The corresponding metric equation is:

$$\text{Compressor cost (in million US\$)} = 2{,}970{,}000 + 1500p \qquad (5.4)$$

where p is the compressor power in kW.

The installation costs consist of material, labor, miscellaneous items, and land costs. The cost breakdown is shown in figure 5–5.

The figure illustrates that the cost of the materials and labor account for 85% of the totals, with land costs and miscellaneous items accounting for the rest.

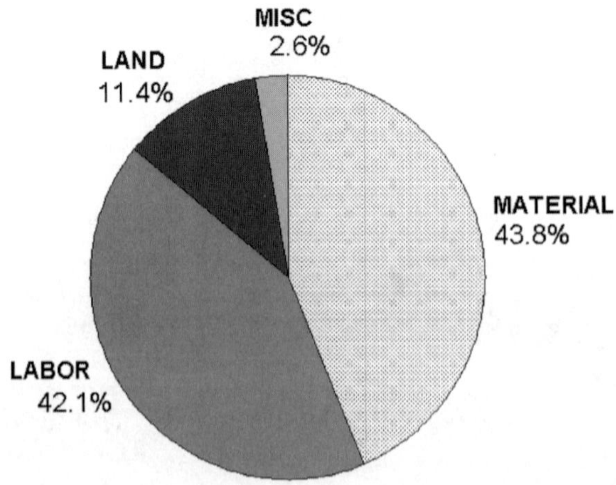

Fig. 5–5. Compressor cost breakdown

Estimation of transmission tariffs—case study

Since transport costs and tariffs can add significantly to the end user of gas, it is of interest to estimate the cost for operating a gas transmission pipeline. Using the data presented above for the cost of pipelines and compressors, it is possible to estimate the carriage cost of gas for a hypothetical 1,000-km gas pipeline. The approach follows the methodology adopted for gas utilization with the following differences:

- No working capital.
- Return on capital investment is set at a discounted cash flow (DCF) rate of 7.5%; using the methodology set out in appendix F for a 2-year construction period, this generates a capital return of 10.9%. This reflects the attitude of many regulatory authorities who regard pipeline systems as long-life "public-good" infrastructure, requiring a lower capital return on investment.
- Pipeline operates for 20 years with a 2-year period of construction.
- The noncapital operating cost is set at 5% of the total capital cost. This covers all outgoing costs on the line (labor, maintenance, gas used in compression, etc.).

A pipeline carrying 200 PJ/y gas would have the statistics given in table 5–2.

Table 5–2. Statistics for hypothetical gas pipeline

Distance	km	1000
Amount moved	PJ/a	200
	MMcf/d	500
Pipeline Costs		
Pipeline (ins)	d	24
Capex/km	US$/km	880962
Pipeline Capex	$MM	880.962
Compression Costs		
Distance between compression	km	50
Number of stations		20
Power Required/Station	MW	3.544
cost/station	$MM	2.975
Compression Capex	$MM	59.506
Total Cost of Line	$MM	940.468
Annual ROC	10.94%	102.887
OPEX (% Capex)	5%	47.023
Annual Costs		149.911
Carriage Cost	$/GJ	0.75

This estimates the cost for carriage of the gas to be \$0.75/GJ. Increasing the return on capital to 14% (representing the more common 10% DCF required by private investors) instead to the 10.9% value results in a rise in the estimated carriage cost to about \$0.90/GJ.

The Alaska Department of Revenue has used a rule of thumb for estimating the pipeline tariff (\$/Mcf):[1]

Cost of the pipeline/
(Total volume of gas carried over the line's lifetime in Mscf)
x 3.35

Using this figure gives a tariff cost of \$0.86/Mcf (about \$0.78/GJ allowing for gas at 1,050 BTU/cf).

Kaufmann and Feizlmayr have detailed the optimization of the pipeline diameter for the transmission for large volumes of gas from the Caspian Sea to China.[2] They also compared alternatives for LNG transport. Although the capital charges for the pipelines are similar, their unit costs are lower than the costs estimated here.

LPG transport by pipelines. Of note is that gas pipelines can be used to transport LPG. LPG is left in the gas stream in excess of that required by the end users of the gas. The excess is stripped out of the gas further down the line in a straddle plant (discussed in chapter 3). Straddle plants have been built in Canada and Western Australia.[3]

Shipping Gas as Liquefied Natural Gas (LNG)

Natural gas can be liquefied at –165°C and shipped in specially built cryogenic ships. Because of the nature of the technology, this method is best suited for the transport of very large quantities of energy (> 200 PJ/y). The capital cost associated with LNG production is detailed in chapter 11. This section will review the cost of transoceanic shipping costs.

LNG tankers are usually constructed and owned by specific LNG project operators. The cost of shipping gas and petroleum liquids is extensively discussed by Masseron.[4] The data presented in that work have been used to develop an operating cost breakdown for shipping. This is summarized in table 5–3 for a carrier of 150,000-m^3 capacity of LNG (about 69,000 t).

The carrying capacity of the LNG tanker has been reduced by 5% to allow for fuelling the tanker from boil-off and to make allowance for carrying some LNG to maintain the ship at cryogenic temperatures. The tanker capital is based on a rate of \$1,200/m^3 of capacity.[5] The ship is considered to have a life of 15 years.

The analysis gives an estimate of \$0.52/GJ.

Table 5–3. Statistics for shipping LNG

Ship capacity	cm	150,000
Actual Capacity	cm	142,500
	DWT	65,550
	GJ	3,565,806
Logistics		
Days/year		350
One way distance	km	5,000
Speed	knots	18
Sailing time	days	6.25
Turnaround time	h	36
One way trips/year		45.16
Sailing days/year		282.25
Port calls/year		45.16
Days in port/year		67.75
Capital Costs		
Capital cost	MM$	171
ROC (15y, 10% DCF)	%	15.19%
Capital costs		25.97
Operating Costs		
Labor	MM$/a	2.52
Fuel	t/day	boil off
Fuel costs	MM$/a	0
Port fees/station	$	80,000
Port charges	MM$/a	3.61
Maintenance	% Capex	4%
	MM$/a	6.84
Insurances	% Opex	15%
	MM$/a	1.95
Misc (victualing etc)	% Opex	10%
	MM$/a	1.3
Total OPEX	MM$/a	16.21
Total Costs	MM$/a	42.18
Quantity Shipped	t/a	1,480,257
	PJ/a	81.41
Shipping Cost	$/t	28.5
	$/GJ	0.52

The variation in the cost of shipping LNG with distance is illustrated in figure 5–6 using the salient details of the ship given in table 5–3. The trend line in the figure is very similar to the data reported by the EIA in its global LNG study of $0.05/MMBTU, with a distance charge of $0.00015/mile.[6] Points in the figure are added to indicate EIA's estimates for moving gas from the areas shown to the U.S. West Coast.

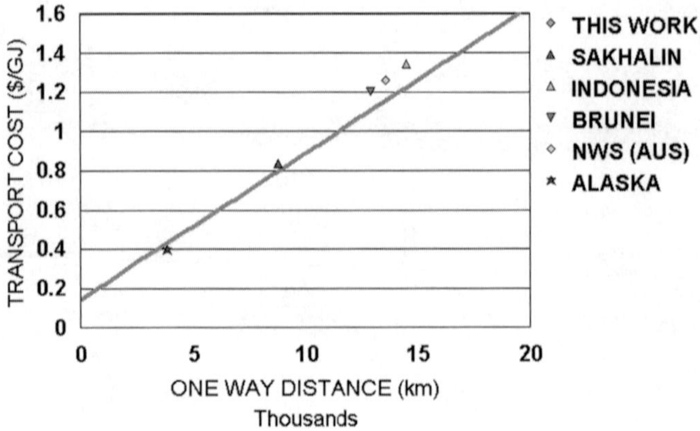

Fig. 5–6. LNG transport cost

Figure 5–6 illustrates that for a 5,000-km one-way distance, the cost of transporting LNG is about $0.52/GJ, and for a 10,000-km distance, the cost is about $0.90/GJ.

Compressed natural gas. Because of the high cost of making and transporting LNG, some groups have recently shown an interest in the use of CNG as a means of transporting large volumes of gas over short (< 1,000-km) distances. CNG is well-known as an alternative transport fuel for vehicles. The concept is similar to this, but on a larger scale, using specialist ships to store the CNG. The advantage for CNG lies in the lower cost of gas clean up, compression, and off-loading facilities, rather than in lower shipping costs relative to LNG. The cost of shipping by CNG is considerably higher than that for LNG at distances greater than 1,000 km. Shipping CNG is discussed in chapter 11.

Transport by hydrates. There has also been some recent interest in the use of gas hydrates as an alternative means of transporting gas. The technology is to form a hydrate at a suitable temperature and pressure and then to transport this as a solid (via trucks, ships, etc.). At the receiving end, the temperature and pressure are adjusted to separate the gas and water. Hydrate transport is claimed to be cheaper than LNG and GTL for long distances.[7]

Shipping Methanol

The first two sections illustrate the capital intensity of moving natural gas either through pipelines or as LNG. An alternative might be to convert the natural gas at the wellhead (or as near as possible) to a liquid such as methanol. The liquid could then be transported by means of a conventional tanker at a cost typically of $1.50/t per 1,000 km. Because methanol is a liquid at ambient temperatures and pressures, this also would have the advantage of saving storage and handling costs at each end of the journey.

For liquids such as methanol, the typical capital is about $350 million for a production rate of about 2,500 t/d (800,000 t/y; about 20 PJ/y). Such technologies are probably best suited to the transport of smaller amounts (10–100 PJ/y) of energy. The cost of production is discussed more fully in chapter 8.

Using a conventional plant, there is little scope for cost reduction by scaling up. A larger scale plant will comprise simple multiples of the smaller (2,500 t/d) plant. Thus, as plants rise in size, the cost per gigajoule will fall, but not by a great deal. This means that any advantage methanol has for the transport of low levels of energy will be lost to pipeline and LNG transportation as the quantity of energy to be delivered rises.

Liquid production also shows other disadvantages. One is the inevitable energy loss incurred in the production of the liquid. For instance, the theoretical efficiency for methanol production is 83%, but in most cases, it is about 73% or less. For the production of liquids by the FT process, the energy loss is higher, with the efficiency of the best cases being about 65%. Such energy losses become increasingly important as the cost of developing the gas field rises.

Methanol also has a major disadvantage in its low calorific value (22.7 GJ/t) when compared to other liquids such as gasoline (46.9 GJ/t). This means that a tanker carrying methanol will transport only half of the energy that it would if it were carrying a hydrocarbon such as crude oil or gasoline.

Following the methodology used in this work, two options for methanol transport are considered. The cost basis is based on studies for the transport of methanol as an alternative to gasoline performed by U.S. Department of Energy.[8] These include a conventional tanker of 40,000 t and a very large tanker of 250,000 t. The statistics are given in table 5–4 and the variation in shipping cost with distance is illustrated in figure 5–7.

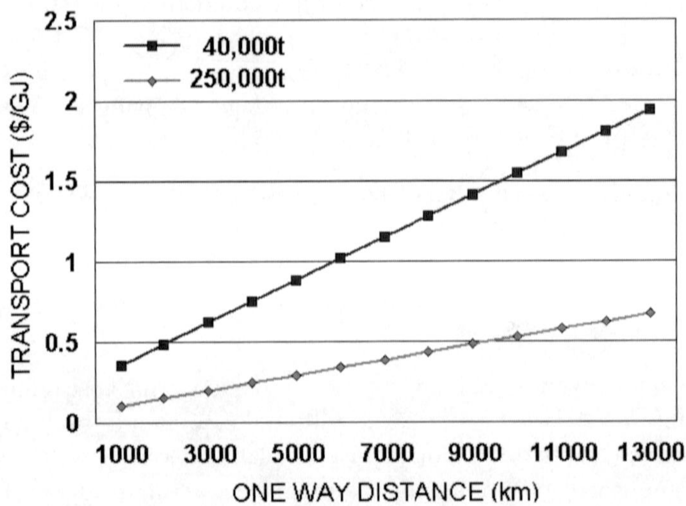

Fig. 5–7. Methanol shipping costs

Figure 5–7 shows that using conventional tankers to transport methanol a distance of 5,000 km costs about $0.89/GJ; about 50% more than the cost of LNG at $0.52/GJ. However, if a project is sufficiently large to warrant the use of large cargo carriers, this could be reduced to $0.30/GJ,or about half the carriage cost of LNG. The use of the very large carrier overcomes the disadvantage of the low calorific value of methanol.

Table 5–4. Statistics for shipping methanol

Ship Capacity	DWT	40,000	250,000
	GJ	908,000	5,675,000
Logistics			
Days/year		350	350
One way distance	km	5000	5000
Speed	knots	12	12
Sailing time	days	9.37	9.37
Turnaround time	h	24	24
One way trips/year		33.74	33.74
Sailing days/year		316.26	316.26
Port calls/year		33.74	33.74
Days in port/year		33.74	33.74
Capital Costs			
Capital cost	MM$	33.31	93.88
Capital costs	MM$	5.06	14.26
ROC (15y, 10% DCF)	%	15.19%	15.19%
Operating Costs			
Labor	MM$/a	2.52	3.31
Fuel	t/day	20	30
Fuel costs	MM$/a	0.95	1.42
Port fees/station	$	60,000	80,000
Port charges	MM$/a	2.02	2.7
Maintenance	% Capex	4%	4%
	MM$/a	1.33	3.76
Insurances	% Opex	15%	15%
	MM$/a	1.02	1.68
Misc (victualing etc)	% Opex	10%	10%
	MM$/a	0.68	1.12
Total OPEX	MM$/a	8.53	13.99
TOTAL COSTS	MM$/a	13.59	28.24
Quantity Shipped	t/a	674,748	4,217,172
	PJ/a	15.3	95.7
Shipping Cost	$/t	20.137	6.697
	$/GJ	0.89	0.30

Methanol quality. If energy is the only consideration, then methanol shipping costs can be reduced to some extent by producing a crude methanol containing higher alcohols. It is relatively straightforward to include about 4% higher alcohols in methanol synthesis (mainly butanols). This boosts the calorific value of the fuel and hence lowers the transport cost in terms of dollars per unit of energy.

DME Transport

An option to transport the energy of gas is to convert the gas into DME. DME is gaseous at ambient temperatures and pressures but is much easier to liquefy that natural gas. DME can be made from gas in much the same way as methanol. DME has a higher heating value (31 GJ/t) than methanol but requires pressurizing or cooling for transport. Transport is therefore similar to that of LPG or ammonia, which can be transported in either pressurized or refrigerated vessels. Costs are intermediate between liquids and specialized LNG carriers.

LPG shipping costs are seasonal and dependent on the business cycle. Typical costs for the spot carriage cost of LPG cargoes are illustrated in figure 5–8 for three sizes of carrier.

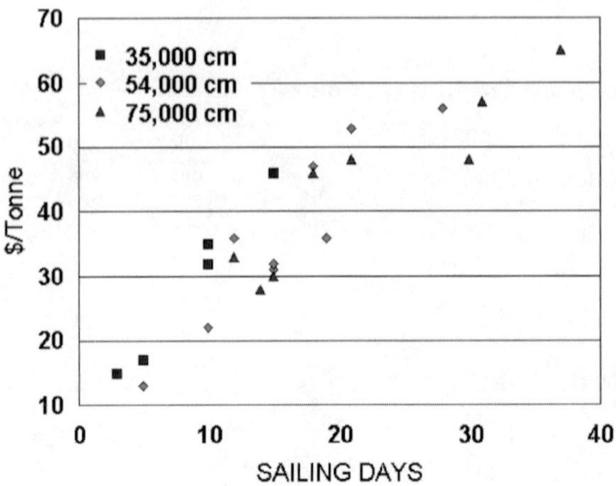

Fig. 5–8. LPG shipping costs. Details from Commercial Services Company Ltd. "Waterbourne LPG," Report, Houston, TX, June 21, 2001.

This figure shows that as the capacity of the ship rises, the spot cargo cost falls for a given distance. Interestingly, the cost variation can be accounted for by scaling with an exponent of 0.65; namely:

$$\text{Cost}[1]/\text{Cost}[2] = \{\text{Capacity}[1]/\text{Capacity}[2]\}^{0.65} \tag{5.5}$$

The result of this normalization is illustrated in figure 5–9.

The trendline has the formula:

$$\text{Transport Cost (\$/t)} = 1.64^{*}\text{Sailing Days} + 4.21 \tag{5.6}$$

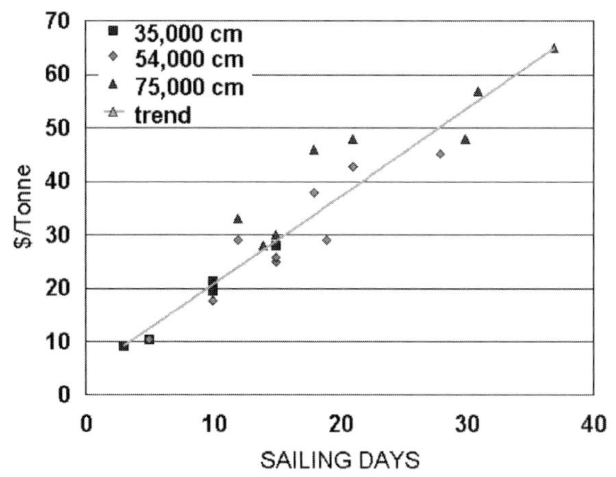

Fig. 5–9. LPG shipping costs normalized to 75,000 m^3

For comparative purposes, these equations can be translated into sailing distances and into energy units ($/GJ) for the liquefied gases of interest. The resulting correlation is shown in figure 5–10 for DME.

This correlation indicates that for a 5,000-km journey, the DME carriage cost (normalized to a 75,000-m^3 ship) would be $0.63/GJ; slightly higher than the cost for LNG, but lower that the estimated cost for methanol.

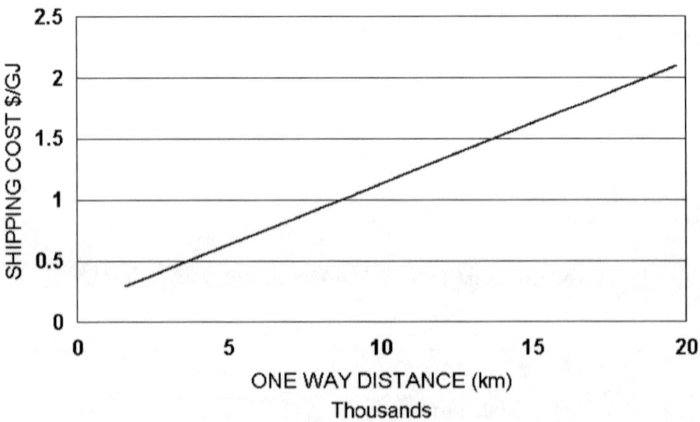

Fig. 5–10. DME shipping costs

Comparative Case Studies

The difference in transport costs can be the critical factor in determining the route by which a gas reserve is developed, particularly for reserves remote from major markets. Thus there is a continuing interest in comparative transport costs.

Alternative liquid products

As developed previously, the transport costs of alternative liquid products are shown in figure 5–11.

The graph clearly illustrates that the cost of conveying energy in the form of LNG in purpose-built high-speed carriers is lower than using methanol in smaller chemical tankers or DME in pressurized or cryogenic tankers. However, for a suitably sized development, shipping methanol in very large carriers would have a lower transport cost than LNG.

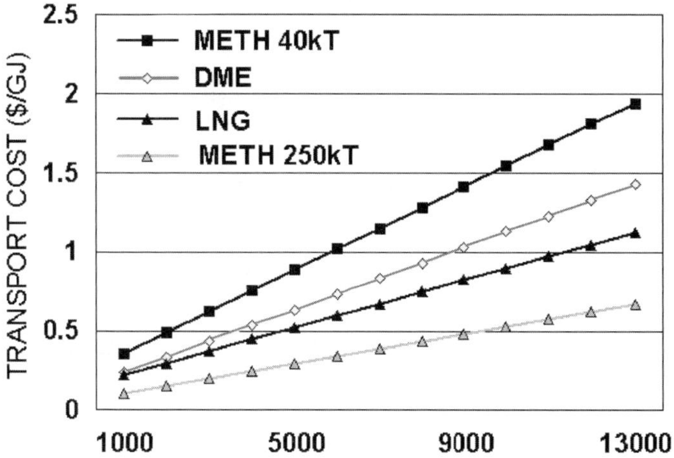

Fig. 5–11. Comparison of shipping costs

Pipelines versus liquids

Another issue for consideration is how the cost of transporting gas via pipeline compares with the liquefying and shipping costs for LNG or methanol. The analysis discussed here gives the results illustrated in figure 5–12.

This simple analysis indicates the competitive position of liquids over pipelines for transporting energy over long distances. For distances less than 1,000 km, pipelines are the most cost-effective solution. One should recall that this analysis does not consider gas conversion costs to producing LNG or methanol, which are dealt with in later chapters. The competitive position of pipelines as a means of moving energy is greatly improved when these costs are taken into account.

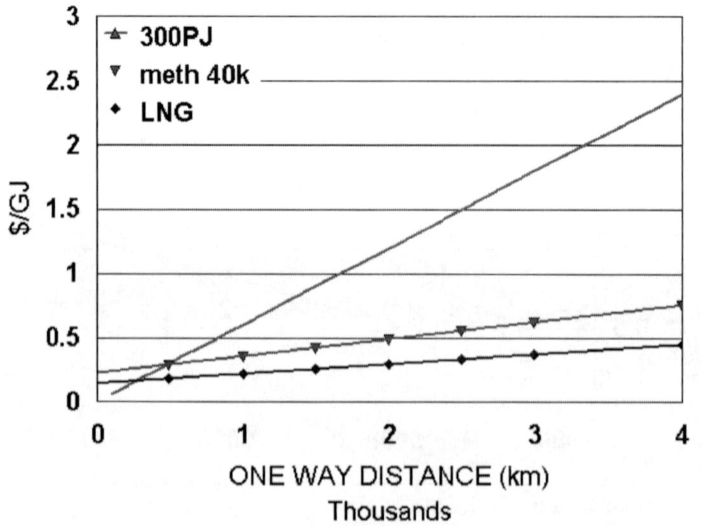

Fig. 5–12. Comparison of pipeline cost to shipping cost

References

1. Quoted Alaska Resources Board method by E. P. Robertson. "Options for Gas to Liquids Technology in Alaska," INEEL/EXT-99-01023 for U.S. DOE DE-AC07-99ID13727; 1999.

2. Kaufmann, K.-D., and A. H. Feizlmayr. "Analysis Pegs Pipeline Ahead of LNG for Caspian Area Gas to China," *Oil & Gas Journal*, March 8, 2004, p. 58.

3. Hawkins, D. J. "Alberta Gas Processing Shows Significant Development," *Oil & Gas Journal*, December 16, 2002, p. 46.

4. Masseron, J. *Petroleum Economics,* fourth edition, Editions Technip, 1990.

5. Kaplan, A., et al. *Oil & Gas Journal*, Wall Chart "2003 LNG World Trade and Technology," quoting Cotton and Co. estimates for 2000–2007.

6. Neil, C. "Distance Continues to Drive LNG Costs for U.S. Delivery," *Oil & Gas Journal*, May 17, 2004, p. 56, quoting UA EIA, "The Global Liquefied Natural Gas Market, Status and Outlook," DOE/EIA–0637, December 2003.

7. Gudmundsson, J. S., et al. "Hydrate Concept for Capturing Associated Gas," SPE European Petroleum Conference, The Hague, October 20–22, 1998; abridged as "Natural Gas Hydrates: A New Gas Transportation Form," in *Journal of Petroleum Technology*, April 1999, p. 66.

8. "Assessment of Costs and Benefits of Flexible and Alternative Fuel Use in the U.S. Transportation Sector—Technical Report Three: Methanol Production and Transportation Costs," U.S. Department of Energy, DOE/PE-0093, November 1989.

6

Power Generation and Thermal Uses of Gas

This chapter will consider the use of gas for power generation and general heating duties. It will primarily address the use of gas for heating duty in medium and large industrial enterprises. Such enterprises can be supplied by gas directly from a specific gas field, and there can be a deal of flexibility in the specifications of the gas supplied.

For most small users, heating is the primary use of gas. For smaller scale operations, such as less than 1 PJ/y, supply is usually via a utility that supplies many end users through a reticulation pipeline network. For these uses, the gas will have been supplied to the utility at an agreed specification. The nature of these specifications will be addressed first.

Pipeline Specifications

Natural gas transmitted through a pipeline has to comply with an agreed specification. The variations in specification in various regions of the world were presented in chapter 3. Here specifications are discussed from the perspective of the end user.

The required user specification is determined by two factors:

- **Maintaining the integrity of the pipeline.** This concerns minimizing those components leading to corrosion or other hazards within the pipeline. A principal concern is the formation of acidic solutions. These solutions would promote corrosion by minimizing the amount of water and the acid gases, hydrogen sulfide and carbon dioxide. Because of its high toxicity, hydrogen sulfide is removed to < 1 ppm. Carbon dioxide is reduced to typically below 2% (by volume). Water control is by dew-point control, which is typically less than 10°C. Heavy hydrocarbons are also controlled by dew point (typically below 0°C) to prevent condensation. This prevents partial blockage and two-phase flow within the line.

- **Maintaining fuel interchangeability.** The fuel specifications allow an interchangeable quality of the fuel for the safe operation of downstream appliances. This is primarily of concern to the end user.

Fuel interchangeability

Only gases with relatively close heating values can be considered interchangeable. Appliance manufacturers generally group gaseous fuels into three groups:

1. Town gas (coal gas), with an HHV typically of 12 MJ/m^3 (about 320 BTU/cf)
2. Natural gases, with an HHV of about 37 MJ/m^3 (about 1,000 BTU/cf)
3. LPG (propane and butane), with an HHV of 94–120 MJ/m^3 (about 2,500–3,300 BTU/cf)

Two gases are completely interchangeable if they give identical flames from a given burner without the need to modify the way it is regulated or change its geometry. Two gases are completely interchangeable if there is no change to:

- The rate of heat flow
- The rate of primary aeration
- The concentration of carbon monoxide and carbon dioxide in the product gas, and if the same yellow points are maintained
- The flame stability with respect to blowout or flashback
- The facility for mutual lighting by various parts of the burner
- The combustion efficiency
- The oxidizing power and the dew point of the combustion products

All of these conditions are rarely met, and in practice, a certain tolerance is permitted. Practically speaking, two gases are interchangeable if they have the same Wobbe Index and the same combustion index.

Wobbe Index

The Wobbe Index determines the limits in the rates of aeration, affecting the height of the blue flame cones. The heat flux of a burner is proportional to the Wobbe Index (W), which is defined as:

$$W = H/\sqrt{\delta} \tag{6.1}$$

where H is the heat of combustion per unit volume, and δ is the specific gravity of the gas relative to air.

There are two indices, corresponding to the higher and lower heating values of the gas. The Wobbe Index is thus influenced by any components in the gas that will significantly alter either of the following:

- **The heating value.** On one hand, this can be the level of inert components (nitrogen, carbon dioxide, and water). On the other hand, this can be components of significantly higher heating value than methane (propane and butane).
- **The specific gravity of the gas.** Methane (the principal component of natural gas) has a low specific gravity. Consequently, this means all of the concentration of the components other than methane, especially those components of high molecular weight and gravity, such as carbon dioxide.

The specification of Wobbe Index thus profoundly influences the required gas composition.

Principal components of gas

The general components of wellhead gas are discussed in chapter 3. Presented here are data for gas relevant to the specification of a gas for pipeline transmission or reticulation through a utility distribution system. Specifications are drawn up along the principles outlined above and can vary over a range of values. Typical specifications are given in table 6–1.

Table 6–1. Typical pipeline gas specification

Component	Unit	A	B
Maximum carbon dioxide	mol.%	3.6	4
Maximum inert gases	mol.%	5.5	6
Minimum HHV	MJ/m^3	37.3	37.3
Maximum HHV	MJ/m^3	42.3	42.3
Minimum Wobbe Index	MJ/m^3	47.3	47.3
Maximum Wobbe Index	MJ/m^3	51	51
Maximum sulfur unodorized gas	mg/m^3	10	10
Maximum sulfur odorized gas	mg/m^3		20
Maximum hydrogen sulphide	mg/m^3	2	2
Maximum oxygen	MJ/m^3	0.2	0.2
Maximum water	mg/m^3	48	48
Hydrocarbon dewpoint over pressure range 2.5 to 8.72MPa	C	<0	<0
Maximum radioactive component	Bq/m^3	600	600
Minimum extractable LPG	t/TJ	1.45	

This table illustrates specifications for two gas types, A and B, which may use the same pipeline system. Type A is rich in LPG and is an example of a gas typically carried by pipeline. For gas of this specification, a straddle plant extracts the LPG at the end of the line. Type B is for gas that is transported through the line but is not sent to the straddle plant. The properties of the key components are given in table 6–2.

Table 6–2. Properties of some key components of natural gas

	MW	HHV kcal/ mole	LHV kcal/ mole	Gross BTU/cf	Net BTU/cf	HHV MJ/ cm	LHV MJ/ cm	sg Air =1.000
Helium	4	0	0	0	0	0	0	
Hydrogen	2	68.3	57.8	325	275	12.1	10.2	
Carbon monoxide	28	67.6	67.6	322	322	12	12	
Carbon dioxide	44	0	0	0	0	0	0	1.519
Nitrogen	28	0	0	0	0	0	0	
Methane	16	212.8	191.8	1013	913	37.7	34	0.553
Ethane	30.1	372.8	341.3	1792	1641	66.7	61	1.038
Propane	44.1	530.6	488.5	2590	2385	93.9	86.4	1.522
n-Butane	58.1	688	635.4	3370	3113	121.7	112.4	2.006
iso-Butane	58.1	686.3	633.7			121.4	112.1	2.006
n-Pentane	72.2	845.2	782	4016	3709	149.6	138.4	2.491
iso-Pentane	72.2	843.2	780.1			149.2	138	2.491
Hexane	86.2	1002.6	928.9	4616	4309	177.4	164.4	
Heptanes +	100.2	1160	1075.9	5216	4909	205.3	190.4	
Ethylene	28.1	337.2	316.2	1614	1513	60.1	56.4	
Propylene	42.1	492	460.4	2336	2186	87.1	81.5	
Butenes	56.1	649.8	607.7	3084	2885	115	107.5	
Pentenes	70.1	806.9	754.3	3836	3686	142.8	133.5	

Competitive Pricing for Small Users

As discussed in chapter 4, for many small uses and appliances, gas competes against light fuel oil, LPG (mainly propane), coal, and electricity. For many end users, after consideration of differences in efficiency and change of burner costs, fuel substitution is based on price. Note that, for the majority of the applications, fuels should be substituted on the basis of the LHV. This effectively increases the gas price by typically 13%–15% more than the often-quoted HHV prices.

Gas price equivalents

In different parts of the world gas is sold (usually on an HHV basis) using different units. Approximate conversion factors for gas are given in table 6–3.

Table 6–3. Conversion factors for gas

$/GJ	$/MMBTU	$/Mscf	cents/cm	c/Therm
1.000	1.055	1.108	3.920	10.551
0.948	1.000	1.050	3.715	10.000
0.903	0.952	1.000	3.538	9.523
0.255	0.269	0.283	1.000	2.692
0.095	0.100	0.105	0.372	1.000

HHV 39.2MJ/cm (1050BTU/cf)

Gas versus light fuel oil

In the urban industrial markets, gas competes against the heavier distillate fuels or the lighter fuel oils (often referred to as industrial distillate fuel or number 2 distillate fuel). Note these lighter fuel oils are sold on a volume basis and shipped in road tankers to end users. Efficiency differences are small, but gas is slightly more cost efficient because local storage facilities are not required. Table 6–4 gives approximate price conversion factors for light fuel oil.

Table 6–4. Price conversion factors for gas and light fuel oil

$/GJ	$/t	$/bbl	cents/gal	cents/L
1.000	44.100	6.029	14.356	3.793
0.023	1.000	0.137	0.326	0.086
0.166	7.314	1.000	2.381	0.629
0.070	3.072	0.420	1.000	0.264
0.264	11.627	1.590	3.785	1.000

(fuel oil density = 860kg/m3, HHV = 44.1GJ/t)

Gas versus LPG (propane)

The cost of transport and storage of LPG is significantly more than that for fuel oil. When gas is available, this tips the balance in favor of gas. Table 6–5 indicates LPG (propane) price equivalents.

Table 6–5. Price conversion factors for propane

$/GJ	$/t	$/bbl	cents/gal	cents/L
1.000	49.600	4.003	9.532	2.518
0.020	1.000	0.081	0.192	0.051
0.250	12.390	1.000	2.381	0.629
0.105	5.204	0.420	1.000	0.264
0.397	19.696	1.590	3.785	1.000

(density = 1969.7L/t; HHV = 49.6GJ/t)

Gas versus coal and electricity

Coal usually enters the smaller appliance market via its derivative, electricity. Gas versus electricity is a complex issue made more complicated by time-varying electricity tariffs. Efficiency considerations are also important.

On an energy equivalent basis, electricity is very expensive relative to all other fuels, but for many end uses, it remains the fuel of choice. Table 6–6 and table 6–7 give the price conversion factors of coal and electricity showing the relatively high price of electrical power. For many applications, power use is almost 100% efficient, whereas other fuels often have a thermal efficiency below 50%.

Table 6–6. Price conversion for coal

$/GJ	$/tonne	$/ton
1.000	27.000	24.494
0.037	1.000	0.907
0.041	1.102	1.000

(HV = 27GJ/t)

Table 6–7. Price conversion factors for electricity

$/GJ	c/kWh
1.000	0.360
2.778	1.000

Power Generation

The modern method for power generation from gas uses gas turbine generators. There are three general schemes: open-cycle gas turbine, combined-cycle gas turbine, and cogeneration. Older plants require steam to be raised by fuel combustion in a boiler. The use for gas in this duty is discussed.

Gas turbines

The gas turbine system is shown schematically in figure 6–1.

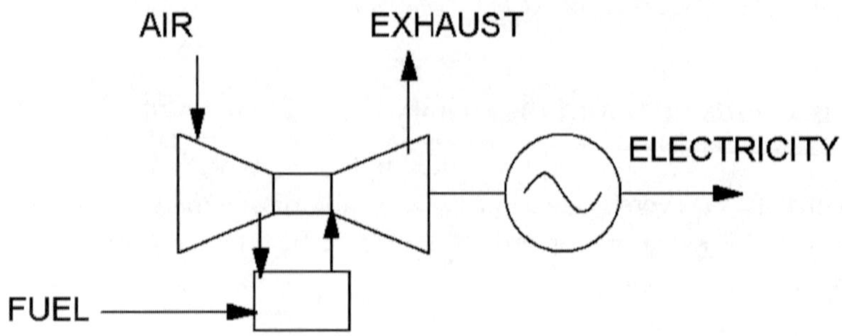

Fig. 6–1. Single-cycle gas turbine

Gas turbines are based on modern aircraft engine technology. Air is compressed in the front end of the unit and passes to the combustion chamber, where fuel is injected. The expanding gases drive the expansion turbine coupled to the inlet compressor and the outlet power shaft, to which the generator is connected. Exhaust gases are discharged to the atmosphere. Since gases are exhausted to the atmosphere, the LHV is relevant in determining thermal efficiency.

Electricity generation using such plants is either referred to as single-cycle gas turbine or open-cycle gas turbine generation. They are relatively cheap and can be brought on-line quickly. They are most often used for generation of small amounts of electricity (< 50 MW) but can be used in the range of 100 MW or more to provide electricity at peak periods of demand.

For these plants, thermal efficiency (energy produced divided by energy in the fuel) is typically 30%–35%. Newer technologies may produce efficiencies of 40%. There are claims that the next generation single-cycle gas turbines will have thermal efficiencies approaching 45%.

Thermal efficiency is improved if the inlet air temperature is minimized. In hot environments, a refrigeration cooler is placed at the air inlet. This requires some power, but the trade-off is usually beneficial in equatorial climates.

Many gas turbines are capable of using any fuel, which can be delivered as a gas or volatile liquid (gas, LPG, naphtha, or kerosene). There are claims that some turbines can use gas oil.

Power barges. Gas-turbine generator sets are available in a range of sizes as packaged units up to 100 MW. The power barge (typically 80 MW) is one interesting variation. These low-cost units usually consist of two 40-MW turbine generators and ancillary power transformer equipment on a barge ready for immediate hook-up to a suitable gas or other fuel supply.

Combined-cycle turbines

An open-cycle turbine scheme exhausts the effluent gas to the atmosphere, and substantial quantities of heat are lost. This lost heat can be used to raise steam in a heat exchanger after the gas turbine exhaust. This produces a combined-cycle system when the steam is also used to produce electricity. The steam is condensed and returned to steam generation. Schematically the system is shown in figure 6–2.

If water in the exhaust is condensed and the sensible heat reused in raising steam, combined-cycle efficiency is determined from the HHV of the fuel. If the water in the exhaust gas is not condensed, then the energy costs should be determined on an LHV basis.

Combined-cycle gas plants typically have thermal efficiencies of about 45%, and new technologies claim efficiencies of 50% or more. These systems are often used for large-scale baseload power generation of 200 MW or more, when the economies of scale justify the increased capital expenditure.

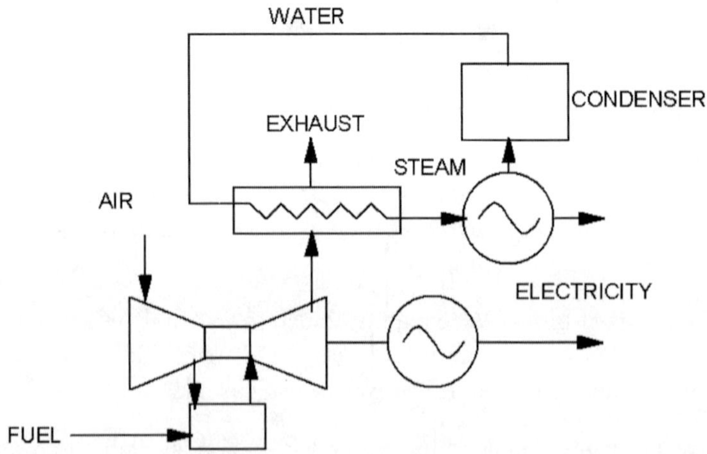

Fig. 6–2. Gas turbine combined-cycle power generation

Cogeneration

If steam is condensed to produce electricity, this results in substantial energy loss, because the heat of condensation will not produce power. If steam can be used in other processes, such as petrochemical operations or paper manufacture, then this produces a higher thermal efficiency overall. This type of scheme (steam export to another activity) is known as cogeneration. Schematically the system is shown in figure 6–3.

In large integrated petrochemical and refinery operations, the boiler feed water is purified to a very high level of purity, and steam is raised to very high pressures. The pressure of the steam is then reduced via steam turbines, thus generating power. The steam, now at lower pressure, is of use to unit operations in the complex. Only excess steam is condensed and recycled within the power generating plant, with the bulk of the condensed steam being returned from the unit operations within the complex. Since all of the available heat is utilized, very high thermal efficiency results. Cogeneration schemes can have thermal efficiencies of 80% or higher.

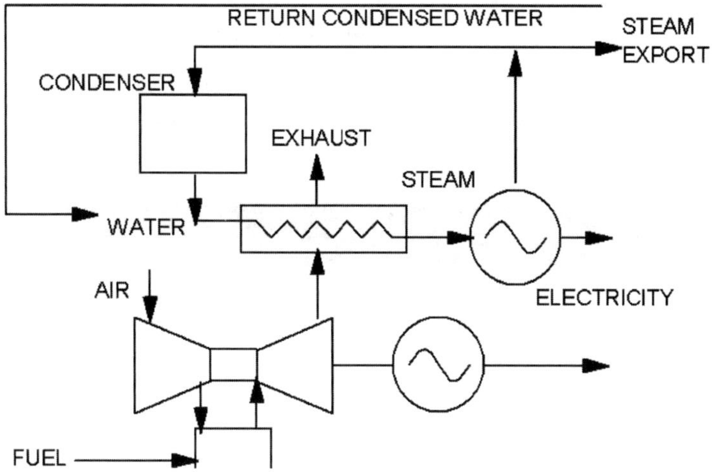

Fig. 6–3. Cogeneration

Steam raising in boilers

The increasing availability of gas makes it feasible to use gas as an alternative fuel in conventional boiler plants, provided appropriate changes are made to the burners. Gas can be fully substituted for an older fuel type, such as heavy fuel oil, or partly substituted, as when gas is used as a supplementary fuel in coal-fired boilers. In most cases, although there is some heat recovery, exhaust gases are rarely cooled sufficiently to condense the water of combustion, so analysis should be on an LHV basis.

Power generation costs

Costs for large-scale base power load generating stations (capacity of 500 MW) are evaluated. Such stations often operate at a load factor (LF) of about 80% capacity.

At this scale of operation, a generating station comprising only open-cycle gas turbines with an operating efficiency of 35% (LHV basis) will consume about 36 PJ/y of gas. On an HHV basis, this is about 40 PJ/y. The capital cost would be about $200 million. A combined-cycle plant of the same capacity would only require about 23 PJ/y gas on an LHV basis, or slightly more than 25 PJ/y on an HHV basis. Although fewer open-cycle gas turbines are required, ancillary equipment of steam generation and water cooling would have raised the cost to about $400 million in 2004.[1]

The salient economic statistics are given in table 6–8.

Table 6–8. Statistics for power generation using gas

		Gas Turbine	Combined Cycle
Capital	$MM	198.25	396.51
Construction period	years	2	3
Return on capital (10% DCF, 15 y life)		15.19%	16.02%
Capital return	$MM/a	30.11	63.54
Operating cost	$MM/a	12.89	25.97
Total non-fuel costs	$MM/a	42.99	25.97
Fuel price	$/GJ	4	4
Efficiency (%)		0.35	0.55
HHV fuel used	PJ/a	39.83	25.34
Fuel costs	$MM/a	159.30	101.37
Total Costs	$MM/a	202.29	190.88
Unit generation cost	c/kWh	5.77	5.45

The efficiencies and gas usage are worked out on an LHV basis. Since gas prices are usually discussed on an HHV basis, the gas usage is recalculated using a 10.5% difference between the LHV and HHV. The open-cycle system is constructed over two years, and the combined cycle, three years. No royalties are assumed payable in calculating the return on capital (ROC) for 10% DCF and a 15-year life. Since the electricity cannot be stored, working capital is ignored.

This example gives the estimate of cost at 5.77¢/kWh for the open-cycle system and 5.44¢/kWh for the combined-cycle system.

Sensitivity to gas price. The sensitivity of the analysis to the price of gas is shown in figure 6–4.

The combined-cycle system has a higher capital cost but uses significantly less fuel. Consequently, the combined cycle plant has lower production costs at higher gas prices. For low gas prices, the simple gas turbine produces a lower cost of generation, since its capital cost is much lower.

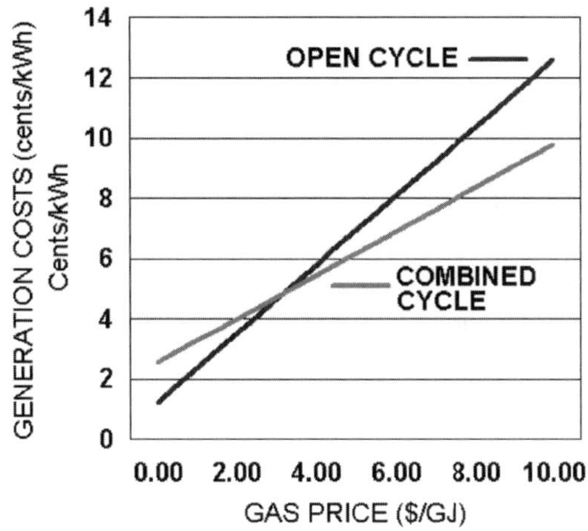

Fig. 6–4. Sensitivity of generation costs to gas price

Sensitivity to the scale of operation. The generation capital costs are sensitive to the scale of operation. If the cost of a plant at one scale of operation is known, an exponential power law is often used to estimate the capital cost of plant at another scale:

$$\text{CAPEX (1)/CAPEX (2)} = [\text{Scale (1)/Scale (2)}]^n \tag{6.3}$$

where

CAPEX is the capital cost of the operation
scale is the operating scale, and
n is a constant.

Values for n are collated and reported for a wide range of unit process operations and full-scale plants in handbooks. They generally range from 0.65 to 0.75 for the processes of interest here.[2] Using a value of 0.70, the sensitivity to the scale of generation can be estimated, and the results are illustrated in figure 6–5.

This figure illustrates that as the scale of the power plant falls, there is a rise in the cost of generation. The rise in generation costs increases dramatically for capacities below 200 MW. Because combined-cycle plants have higher capital costs, the rise in price is more dramatic.

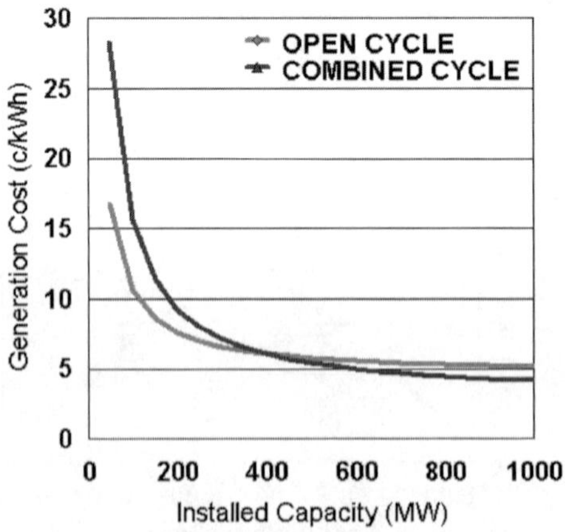

Fig. 6–5. Sensitivity of generation costs to scale

Trade-off between open-cycle and combined-cycle plants and gas price. In both figures 6–4 and 6–5, there is a point of intersection between the open-cycle and combined-cycle generating costs. This intersection represents the position of trade-off between the increased capital cost of adopting a combined cycle and the savings in gas costs. There is interest in determining this trade-off value, since it can be used as a measure for the greenhouse gas savings by moving to the more efficient technology. This point is determined by the variables:

- The gas cost ($/GJ)
- The power output (MW) scale
- The rate of capital recovery
- The scaling factor (typically 0.7)
- The load factor (typically 80%)
- The efficiency factors for open- and combined-cycle plants

Figure 6–6 gives a representative plot for the equivalence point using different levels of power output and gas price. For low gas price and low power output, an open-cycle gas turbine is more economical than using combined-cycle schemes. For either large power requirements or high gas prices, the combined-cycle scheme is more cost-effective.

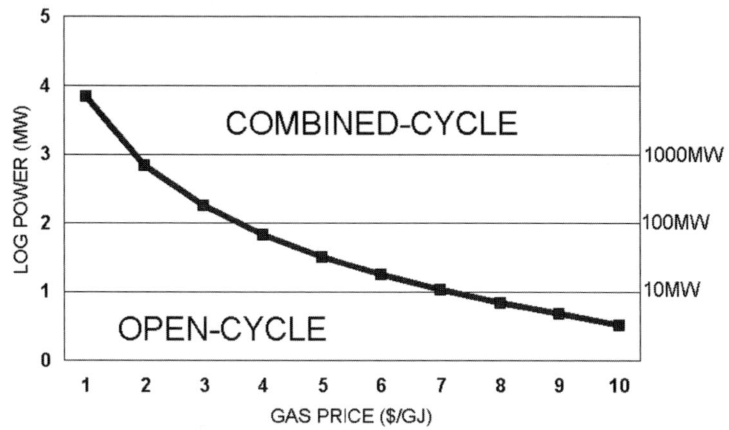

Fig. 6–6. Equivalent unit production costs for open- and combined-cycles plants

Electricity tariffs. The preceding analysis produces an estimate for the production costs of electricity using gas. This is not to be confused with the tariff. Electricity tariffs are determined after several other costs are taken into account by the electricity distributor/utility.[3] For a large industrial user, these are:

- An operating allowance of typically 3%.
- Transmission and distribution losses of typically 8%. (Losses from theft are very high in many parts of the world, possibly as much as 30% of the amount generated.)
- Utility overhead costs of typically 25%.

This produces the tariff for a typical large industrial operation within a normal power distribution network. For very large users (such as aluminum smelters), special rates often apply. Sometimes the tariff for these users is at or below the nominal marginal cost of production, on the grounds that the cost difference will be made up from other uses in the system. This works to the benefit of the power utility by facilitating major investment in large baseload power generation. The cost of production of aluminum is discussed next.

Energy-Intensive Industries

Table 6–9 provides an overview of the use of gas in several industries that require the provision of thermal heat. Note paper and alumina require significant amounts of electric power as well as fuel for heating. Paper and alumina manufacturers benefit from cogeneration schemes. Aluminum is highly energy intensive, and the major portion of the energy is required as electricity. Alumina manufacture is discussed in a later chapter.

Table 6–9. Gas use in some energy intensive industries

Product	Used for	Energy Used (GJ/t)	Notes
Paper	Pulping and drying	5	2/3 used for pulping as power
Glass	Melting	6	2/3 used to maintain temperature of glass at 1500°C
Cement—dry process	Making clinker	3.5	Multiple fuel often used—gas, coal, waste oil
Cement—wet process	Making clinker	5	Multiple fuels often used
Alumina	Drying and calcining	3 to 5	product from Bayer process is calcined to form aluminium oxide
Aluminium	Smelting	55	15% used to heat pot-lines, rest used as power

Paper making: principal unit operations

Figure 6–7 illustrates the principal processes for the production of paper. Wood chips and chemicals enter the process in the pulping plant. This mechanically reduces the chips to fiber. The fibers are passed to the Fourdrinier, which produces the paper. The wet paper is then dried in steam rollers and wound onto the reel.

The plant can be seen to require both mechanical power for pulping operations (about 2/3 of the energy used) and steam for drying. Paper manufacture is ideal for small-scale cogeneration schemes with a capacity of about 20 MW to 50 MW. The mechanical power is provided by electric drives, and steam is used for drying.

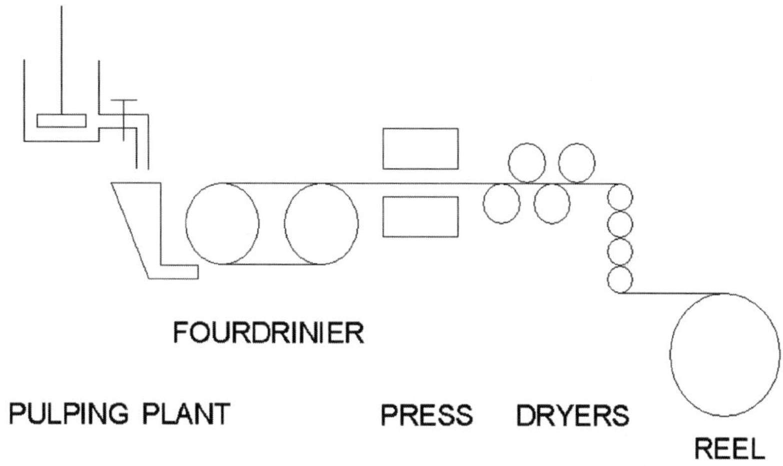

Fig. 6–7. Paper making—principal operations

Cement manufacture

Cement is made by calcining limestone and sand (silica) together with some alumina and ferric oxide to act as a fluxing agent. There are various types depending on the chemical composition. In the dry process, the raw materials are ground together in a dry state. In the wet process, the raw materials are ground with water to produce slurry.

The dry feed or slurry is then passed to a kiln and burned to produce a Portland cement clinker. The kiln is typically a rotary type made of a steel cylinder 150 ft–500 ft in length and 8 ft–16 ft in diameter, lined with refractory brick and slightly inclined to the horizontal.

Fuel in the form of coal, oil, or gas is blown with hot air heated by passing over cooling clinker at the lower end of the kiln. The burning zone is at the front end of the kiln, where temperatures of 1,450°C to 1,600°C are reached.

Early designs were thermally inefficient. There are several variations using heat exchangers and raw material preheating in order to improve this.

From the emissions standpoint, the cement clinker removes metal contaminants of concern. The process is useful in disposal of intractable material contaminated waste, such as sump oil.

Natural gas use in incinerators

There is increasing interest in the disposal of city garbage by incineration. This considerably reduces the volume of garbage going to landfills. In order to aid combustion, gas is often added as a supplementary fuel. Most projects require some form of government support to achieve economic viability.

Garbage has a relatively low heating value (10 GJ/t to 12 GJ/t). The low calorific value results in low combustion temperatures, leading to the production of organochlorines (from PVC and similar polymers). These can be an extreme environmental hazard. In order to increase the temperatures, supplementary fuel (and sometimes oxygen) is added. Typically 1 GJ of additional fuel is added per tonne of garbage (1 MMcf/1,000 t).

This additional heat energy is used to raise steam and so produce export electricity in order to offset the cost of the facility. From the perspective of a power generation plant, the overall efficiency is only about 20%. This means that the facility requires high power prices or other incentives to cover the operating cost.

Economics. A 1,000-t/d facility (1 TJ/d; 0.35 PJ/y of additional gas) would produce 27 MW at a cost of $40 million.

Power Generation for Energy-Intensive Industries

Gas can be used to generate power that is used directly for another (downstream) purpose. Some processes, particularly electrochemical processes, are extremely energy intensive, and given the efficiencies of power generation, vast quantities of gas can be consumed. Two such cases are aluminum smelting and caustic chlorine/polyvinyl chloride (PVC).

Aluminum smelting

The Hall-Heroult process developed in 1886 makes aluminum. Alumina (Bayer process) is mixed with cryolite and other additives and melted in the *pot* at 940°C to 980°C. This heat can be supplied directly by gas. However, a much greater amount of energy is required in the form of electricity to electrolyze the melt and produce the aluminum. The aluminum is produced at carbon cathodes and is drawn off. Oxygen formed at the carbon anodes burns the anodes to carbon oxides. The pot line can comprise up to 100 or more pots.

The power requirement is in the order of 55 GJ/t (13 kWh/kg to 15 kWh/kg of aluminum). If this power is generated from gas, large quantities of gas are required (120 GJ/t). For a typical 200,000 t/y pot line, 25 PJ/y of gas would be required.

Aluminum smelting has been targeted by Middle East gas producers as a means of converting their low-value gas into a high-value exportable commodity.

Aluminum smelting economics. Aluminum smelting operations are very large users of electricity. The amount of power used generally requires major investment in generating facilities, and this impacts the process economics.

Conventional wisdom indicates the smelter economics as summarized in table 6–10.

Table 6–10. Aluminium smelter economics ($/t aluminium)

Alumina cost	375
Anodes	110
Labor	100
Other costs	75
Capital costs	440
Energy	400
Total	1500

Data taken from T. Siglaugsson. "Karahnjukar Hydropower Project," Reykjavik, Iceland, 2002, quoting ACIL consulting work for the Australian Government, also summarized in "Light Metals Action Agenda—Working Paper No. 1: Aluminum," Australian government, August 2001.

Note that these economics are in terms of U.S. dollars per tonne of aluminum, with aluminum at $1,500/t, which was a typical world price in the period 2000–2004. The data have been reanalyzed for a smelter with an output of 200,000 t aluminum, with the following findings:

- Alumina is the basic raw material. Typically the requirement is for 1.89 t alumina for every tonne of aluminum. The cost used is about $200/t. The cost of production of alumina is discussed in detail in a later chapter.
- Carbon anodes are continually consumed in the process at a rate of about 0.45 t carbon for every tonne of aluminum. The anodes are mainly made from petroleum coke bound with coal tar pitch. Petroleum coke is produced from upgrading the residue of heavy crude oils. Its price is determined primarily by its alternative use as a solid fuel and is therefore linked to some extent by the price of coal. Typical prices are about $15–$25/t.[4] The cost for manufacture of anodes is taken as $50/t. To obtain aluminum costs about $110/t, as shown in table 6–10.
- The labor cost of a smelter varies considerably with location. More than 600 people will be employed, either directly or as contractors, for the smelter being considered.
- Other costs include insurance and maintenance materials.

This leaves the return on capital and the electricity price to be considered. The view expressed by the authors for the data in table 6–10 is that smelter costs were about $4,000/t aluminum in 2001. This translates to a cost of about \$5,000/t aluminum in late 2004.[5] For a capital return below \$500/t aluminum, this requires a 30-year, 7.5% DCF rate of return. One can suppose that such a long-term, low DCF is part of the trade-off for very low electricity prices.

A 200,000-t smelter will require about 350 MW of continuous power delivered at high voltage, thereby minimizing transmission costs. This is a major portion of a large baseload power plant. It is often the instigator of a major investment by power utilities, which are generally government or government-backed utilities. Such investments are considered long-term strategic investments, justifying a long-term view (30 years) and a relatively low DCF (5% or 7.5%).

The continuous power demand allows the power generator to operate at a high efficiency with the certainty of a long-term purchase price contract. This justifies the low tariffs required. For the case considered here, the power charge is about 2.5¢/kWh.

Implications for gas price. One can review these gas price/generator cost curves to determine their implications related to gas prices. For this level of tariff, it is apparent that power will have to be generated from large-scale plants, with gas available in the region of \$1/GJ or less.

Caustic chlorine/PVC

The production of PVC requires the production of ethylene and chlorine. Ethylene can be made from the ethane component of natural gas (by gas stripping in a gas plant), and chlorine can be manufactured electrically using electricity produced from natural gas. Chlorine is produced by the electrolysis of brine in a cell. The cell also coproduces caustic soda, which is used in the manufacture of soap and in the production of alumina. This is illustrated in block flow form in figure 6–8.

A gas plant strips ethane from the natural gas. The ethane is passed to the ethylene plant. Methane gas is used to produce power that is used, in turn, to make chlorine (Cl_2) and caustic soda (sodium hydroxide, NaOH). From the ethylene and the chlorine are obtained first ethylene dichloride (EDC), and then vinyl chloride monomer (VCM). Both of these chemicals are traded internationally. PVC is then produced from VCM.

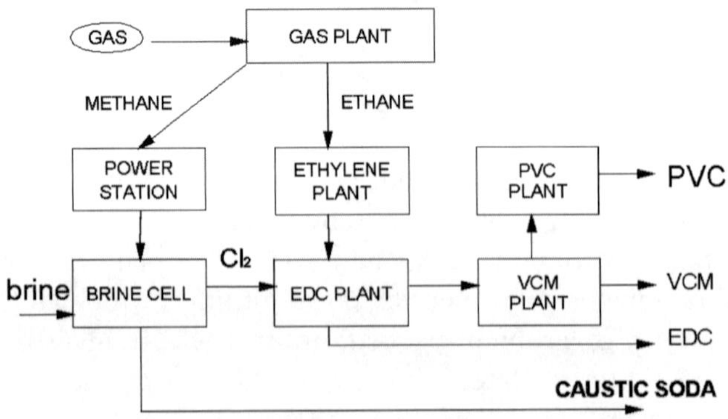

Fig. 6–8. Caustic chlorine and PVC

In the manufacture of chlorine, older mercury cells are now being replaced by diaphragm and membrane cells. The caustic chlorine plants are operated on a wide range of scales from 50,000 t/y chlorine to more than 350,000 t/y chlorine. A 330,000-t/y plant will produce 374,000 t of caustic soda and 580,000 t/y of PVC.

The energy consumption of chlorine plants is high, in the region of 12 GJ/t of chlorine. This is required as electric power. If the energy source is gas using a combined-cycle plant, about 27 GJ/t would be required. There is an advantage in that waste heat can be utilized in the downstream manufacture of EDC and VCM.

One of the key differences between caustic chlorine processes and other electricity-consuming processes is their high level of flexibility. This is especially so with the older mercury cells, which can switch from virtually no power to full power within a few minutes. This permits load-sharing and tariff management throughout the day, achieving very low effective electricity costs.

For example, during the 1970s, ICI PLC operated the very large caustic chlorine plants at Runcorn in the United Kingdom in such a manner. This permitted the site's power generation plant to export power to the United Kingdom's electricity grid during times of high demand and tariff. Conversely, they could import low-priced power during the low-demand, low-tariff times (midnight to 6:00 AM). The mercury cells at Runcorn were said to be able to consume 10% of the United Kingdom's baseload power.

Nitrogen Oxide Emissions

One of the major pollution concerns when using fossil fuels is the production of nitrogen oxides. Nitrogen oxides are oxidizing agents and are major contributors to the formation of photochemical smog. Although nitrogen in the fuel can contribute to the problem, nitrogen oxides are formed when air is exposed to high temperature.

$$N_2 + O_2 = 2NO \tag{6.4}$$

and

$$2NO + O_2 = 2NO_2 \tag{6.5}$$

Although the main nitrogen oxides are nitric oxide (NO) and nitrogen dioxide (NO_2), a small portion of nitrous oxide (N_2O) is also produced. Nitrous oxide is a powerful greenhouse gas, with a greenhouse impact 310 times that of carbon dioxide.

For brevity, the three oxides of nitrogen are commonly referred to as NOX or NO_x.

One of the major problems is that the NO_x-forming reactions are a function of temperature. Increasing temperature favors NO_x formation. The problem for natural gas is its inherently higher flame temperature compared to the fuels it displaces. This can lead to an increase in NO_x when gas displaces another fossil fuel. This is overcome by appropriate adjustments to the burner and furnace operation.

Takacs and others have suggested that the amount of NO_x emitted from any fuel can be quantified using a series of emission factors that are calculated for each pertinent part of the combustion process.[6] These emission factors cover:

- **Fuel type.**
- **Hydrogen content (volume %) of the fuel.** Hydrogen has a very high flame temperature and hence contributes significantly to NO_x.
- **Control technology.** Introducing air or fuel into different parts of the flame can lower the combustion temperature and hence lower NO_x.

- **Flue gas recirculation.** This lowers the flame temperature by diluting the combustion gases and hence reduces NO_x.
- **Air preheating.** This saves energy but also increases the flame temperature and hence increases NO_x.
- **Moisture content of the air.** Higher moisture content lowers the flame temperature and hence NO_x.
- **Burner intensity.** Forced-draft burners have high burner intensities that increase NO_x.
- **Fuel nitrogen.** Nitrogen may be bound in the fuel as ammonia or other nitrogen-containing compounds. This results in reactive nitrogen being delivered to the center of the combustion process, hence increasing NO_x. However, the correlation is not linear. Reactions that convert the bound nitrogen into nitrogen gas also occur. As these processes decrease the effective amount of nitrogen in the feed, they serve to lower the amount of NO_x produced.

The emission factors can be quite complex. They depend upon the nature of the furnace burner, the amount of flue gas being recirculated, the moisture content of the air, and the amount of air preheating.

NO_x from domestic appliances. One of the issues of interest to regulatory authorities charged with pollution control is the quantity of NO_x generated from domestic gas appliances. Domestic appliances rarely have any NO_x mitigation facilities. The base data given by Takacs can be used to approximate NO_x emissions, making an allowance for the humidity of air. This suggests that a city with urban gas reticulation for domestic burners will emit approximately 45 t NO_x/PJ of gas consumed. For a large city with 10 PJ of gas used in the domestic sector, this corresponds to the production of 450 t/y NO_x.

This corresponds to a concentration of about 130 ppm in the flue gas stream, which is broadly similar to the measured values for uncontrolled appliances. Several governments are now directly attacking this issue by bringing in stricter regulations on gas-burning appliances.[7]

References

1. Recent costs for a variety of coal and gas power plants have been collated by A. Matysek, et al. in Australian Bureau of Agriculture and Resource Economics, ABARE E-Report 05.1, January 2005.

2. Perry, R. H., and D. W. Green, editors. *Perry's Chemical Engineers' Handbook,* seventh edition, McGraw-Hill, 1997, tables 9–48 and 9–50.

3. State Electricity Commission of Victoria, 1991/1992 Annual Reports, quoting data from Electricity Association Services Ltd. UK.

4. Thomas, K. "Petroleum Coke to Compete with Coal in Power Plants, Analysts Say," *Oil & Gas Journal Online,* September 19, 2001; also J. Smith. "Analyst Predicts Surge in Petroleum Coke Use in Power Plants.

5. This is consistent with the book value of the Portland Aluminium Smelter, Victoria, Australia, in "Victorian Equity Trust," 1987, inflated to 2004 costs.

6. Takacs, T. J., D. L. Juedes, and I. D. Crane. "Method Estimates NO_X from Combustion Equipment," *Oil & Gas Journal,* June 21, 2004, p. 48.

7. An overview of this subject can be found in B. Joynt and S. Wu. "Nitrogen Oxide Emissions Standards for Domestic Gas Appliances—Background Study," Department of Environment and Heritage, Australia, February 2000.

7

Chemical Intermediates: Synthesis Gas and Hydrogen

From the chemical standpoint, the principal component of natural gas, methane, is inert (apart from burning). Further, from the standpoint of thermodynamics, methane is more stable than many of the products one might wish to make. This means it is very difficult, if not impossible in many cases, to directly convert gas into a more useful product. To overcome this, gas is converted into a more reactive intermediate product, the conversion of which can be controlled to obtain a desired product.

The usual intermediate is known as synthesis gas. This is any mixture of carbon monoxide and hydrogen and can be produced in a variety of ways.[1] Thermodynamically, synthesis gas is accessible from natural gas either by reaction with water at high temperature or partial combustion (reaction with oxygen).

Unlike methane, synthesis gas is very reactive and can be converted by a wide range of processes into a variety of useful chemicals and materials. The production of synthesis gas is central to downstream production of many other chemicals.

Synthesis Gas for Downstream Manufacturing

In principal, synthesis gas can be produced from any carbon source. The most important sources are fossil fuels, such as natural gas. It can also be produced from biocrops or other recyclable materials (garbage), and so it is a means for the sustainable production and recycling of materials. There are two principal methods for producing synthesis gas: steam reforming and partial oxidation.

Depending on the source of the feedstock and the method chosen, the raw synthesis gas is produced over a range of concentrations of carbon monoxide, carbon dioxide, and hydrogen. The ratio of hydrogen to carbon monoxide is known as the stoichiometric ratio (SR) of the gas. This ratio is often used to characterize the synthesis gas produced.

Steam reforming

In steam reforming, the feed (for example, natural gas) reacts with high-temperature steam over a catalyst at about 800°C.

The feed can be any hydrocarbon (CH_x) ranging from methane to naphtha. The key requirement is that the feed must be free of sulfur, which would deleteriously impact the performance of the catalyst.

The principal catalysts are nickel based and operate between 800°C and 900°C. Other metals can be used under special conditions, such as platinum, which can operate at a lower temperature.

The overall reaction is:

$$(a + b + d)CHx + nH_2O = aCO + bCO_2 + cH_2 + dCH_4 \qquad (7.1)$$

This is an equilibrium reaction. The relative proportions of the products (a, b, c, and d in the above equation) are determined by the steam content (n), oxygen in the feed (such as carbon dioxide), and the temperature and pressure. Handbooks and simple computer simulations are widely available that detail the equilibrium over a wide range of feedstocks and conditions.[2]

The steam reforming of natural gas proceeds according to the following partial reaction:

$$(1 + x)CH_4 + H_2O = CO + 3H_2 + xCH_4 \qquad (7.2)$$

Carbon dioxide and additional hydrogen are produced by the WGS reaction (see below).

$$CO + H_2O = CO_2 + H_2 \qquad (7.3)$$

As a consequence of the WGS reaction, carbon dioxide is produced. The result is that the final product comprises carbon monoxide, carbon dioxide, hydrogen, and methane. Methane present in the product is often referred to as *methane slip*.

Natural gas has the advantage over alternative feedstocks, usually having the lowest cost in energy terms, lower plant costs, and low sulfur content. Sulfur content can be deleterious to steam reforming and to many downstream processes. When available, natural gas is used in preference to any alternative fuels.

Steam reformers

Over all practical conditions, this reaction is highly endothermic; that is, it requires the input of large amounts of heat. The method of heat generation, transfer, and heat exchange and the handling of waste heat are central to the efficient operation of steam reformers. The differences in approach characterize the various proprietary types of steam reformers. The layout of a typical steam multitubular reformer is shown in figure 7–1.

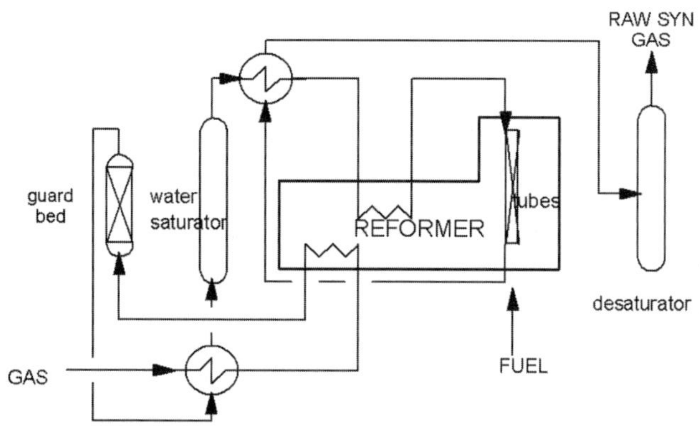

Fig. 7–1. Natural gas steam reformer

Gas is heated and then passes through a guard bed to ensure that no sulfur is present. It is then saturated with water, and the mixture is heated to about 900°C. The gases are then passed over a catalyst, which promotes the reforming reaction. In order to ensure efficient heat input into the catalyst, it is held in tubes of about 10 cm in diameter and about 10 m in length. In a typical steam reformer of this type, thousands of these tubes are held within a heated furnace (sometimes referred to as a *box*). Different process licensors offer differing approaches to heating the tubes and ensuring an even rate of heating over the mass of tubes.

In adiabatic reformers, the heated gas feed is passed over a single packed bed. The endothermic reaction removes the heat from the system, and the gases exit the reformer at a lower temperature. Adiabatic reformers generally operate at lower pass conversion than tubular reformers, but this is acceptable for some applications.

After passing through the reforming section (tubes or adiabatic converter), the gases are cooled to below the water dew point. The excess water is removed and recycled to the inlet water saturator.

Although the flow lines are complex, the equipment is simple but large in size. No input other than water and gas is required.

Steam reforming produces a gas of relatively high carbon dioxide content. In addition, there is often several percent slippage of methane through the reactor (d in eq. 7.1 above). The product gas composition is typically $6H_2$: $2CO$: CO_2, with a methane content of 4% or 5%. The composition can be altered for different end uses by varying levels of steam and by changing the operating pressure and temperature.

Steam reforming is best performed at low pressure, and consequently gas compression downstream is usually required.

Carbon formation. The steam reforming reaction can be considered as the sum of several underlying fundamental processes. One of these reactions is the formation of carbon from reactions such as:

$$CO + H_2 = C + H_2O \tag{7.4}$$

Carbon formation is suppressed by using excess steam and by using additives in the catalyst that promote the gasification of carbon (the reverse of the above reaction).

Carbon deposition is a particular problem with heavier feeds such as LPG or naphtha, and specialist catalysts containing special promoters are used. However, the use of a promoter lowers the steam reforming activity.

For dry natural gas, the catalyst of choice will be low in promoters. However, should heavier material (propane, etc.) be included in the feed, rapid lay-down of carbon can occur.

In the mid-1980s, rapid lay-down of carbon is thought to have occurred in a large steam reformer at the gas-to-methanol-to-gasoline plant in New Zealand. This likely occurred when the gas supply was switched between fields of differing composition. The loss in activity in the reformer due to rapid coking led to overheating of the reformer tubes and consequential melting and destruction of the reformer internals.

Partial oxidation

In partial oxidation, the feed is burned in a restricted amount of oxygen. Any feedstock can be used, including solids. The reaction chemistry is:

$$CH_x + nO_2 = aCO + bCO_2 + cH_2 + dH_2O + eCH_4 + fC \quad (7.5)$$

The reaction reaches a position of thermodynamic equilibrium in a flame. Rapid cooling (quenching) obtains the required synthesis gas. The algorithms for predicting the reaction outcomes over any condition of feed type, temperature, pressure, and oxygen content have been published by Montgomery and others.[3]

Partial oxidation requires oxygen produced in a separate process operation. Oxygen production is both energy and capital intensive. However, this process has the advantage of small size, ability to operate at high pressures (up to 100 bar), and no requirement for catalysts. Figure 7–2 illustrates the principle of synthesis gas production by partial oxidation of natural gas.

Oxygen and natural gas are separately heated (not shown) and mixed at the burner tip of a combustion lance held within the partial oxidation pressure vessel or gasifier. This is internally lined with firebrick to prevent the flame from impinging on the pressure vessel metal. Water or steam can also be injected in the system at the lance tip.

Synthesis gas is produced in the flame and is either quenched with large volumes of water or more efficiently is cooled in a waste heat boiler (WHB). The WHB raises substantial volumes of steam.

Under good process operation, carbon formation is minimal from natural gas, but it is significant for heavier feeds such as naphtha. A scrubbing tower washes the synthesis gas free of soot. The carbon is removed in a settling tank, and it can be formed into a slurry with water and reinjected into the gasifier for disposal. Water picked up in the scrubber is reduced in a cooler and knockout pot and may be further dried using drying sieves. For heavier feeds, such as naphtha, the wash tower can use naphtha feed to wash out the soot. This is then injected into the gasifier.

The temperature in the flame is very high (typically greater than 1,700°C). This temperature favors the formation of carbon monoxide over carbon dioxide. Consequently, partial oxidation produces a synthesis gas containing very little carbon dioxide with almost no methane slippage. Total elimination of methane can rapidly lead to the formation of an explosive mixture within the gasifier. The measurement of the methane slippage (typically less than 0.1%) is used to control the operation of the unit.

The gas stoichiometric ratio is about 1.8 H_2 : CO. Increasing oxygen input and adding some carbon dioxide can move this value down to about 1, and addition of water can increase the value to about 2.

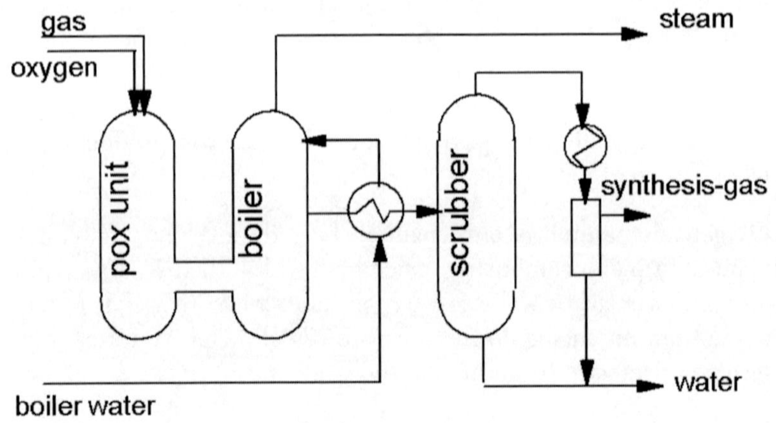

Fig. 7–2. Natural gas partial oxidation

The WGS process

Steam reformers and partial oxidation units are the workhorses of the chemical process industries. There are many hundreds of both steam reformers and partial oxidation units in use. Many downstream operations require a specific synthesis gas stoichiometric ratio. This can be achieved from either process using the WGS reaction:[4]

$$CO + H_2O = CO_2 + H_2 \tag{7.6}$$

This is an equilibrium process, thus driving the reaction to the right will increase the hydrogen content and increase the stoichiometric ratio. Carbon dioxide is easily stripped from the gas in acid gas operations; the Benfield and related processes are often used.[5] Conversely, driving the equilibrium to the left and removing water removes hydrogen and carbon dioxide and lowers the synthesis gas stoichiometric ratio.

Figure 7–3 illustrates the main process operations of the WGS process when low carbon monoxide content is required, as in the production of ammonia or hydrogen.

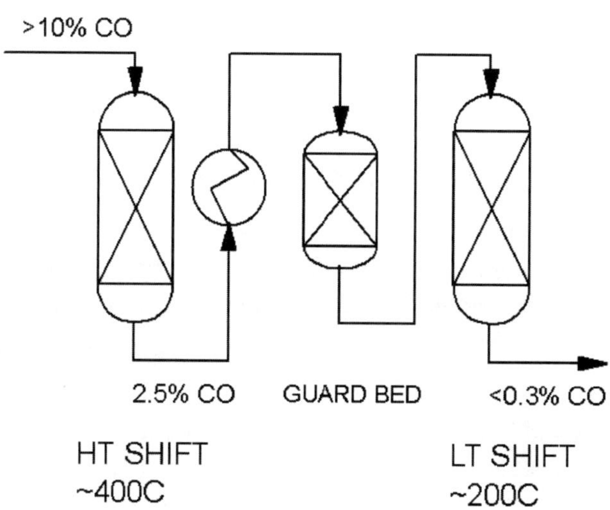

Fig. 7–3. Water gas shift

Gas that is high in carbon monoxide enters the high-temperature shift converter at about 400°C (HT shift). This reduces the carbon monoxide concentration to about 2.5% on a dry basis. Interstage cooling reduces the temperature to about 200°C, and the gas enters the low-temperature reactor (LT shift), where the carbon monoxide is reduced to below 0.3%. The LT shift catalyst can be easily poisoned and is usually preceded by a guard bed. Residual carbon monoxide is removed by absorption in copper solutions, methanation, or by cryogenic separation.

Reverse WGS

The duty of a reverse WGS process is to convert carbon dioxide into carbon monoxide:

$$CO_2 + H_2 = CO + H_2O \tag{7.7}$$

This is achieved in a conventional steam reformer. The process can be integrated into the overall unit operations by extracting carbon dioxide and recycling to the reformer or partial oxidation system. This process results in a synthesis gas of low stoichiometric ratio.

Carbon dioxide recycling and its conversion require a series of highly endothermic reactions. The heat required to drive these processes is best achieved through tubular reformers, where the required additional heat can be easily added externally (increased fuel consumption). This is not as readily available to adiabatic reformers or partial oxidation units, where the amount of carbon dioxide recycling sets limits to the operation.

In practice, high carbon monoxide feed is best achieved using a tubular reformer operated at high temperatures. A typical approach is shown in figure 7–4.

The natural gas enters the steam reformer, which operates at high temperatures and low pressures. The water content of the reformer feed is controlled by steam injection. The methane slippage is relatively high, and the raw gas exits the reformer and passes to a partial oxidation unit. Oxygen is added, and the mixture is burned primarily with the consumption of the methane and some hydrogen. Carbon dioxide produced in the reformer is then extracted and recycled back to the reformer.

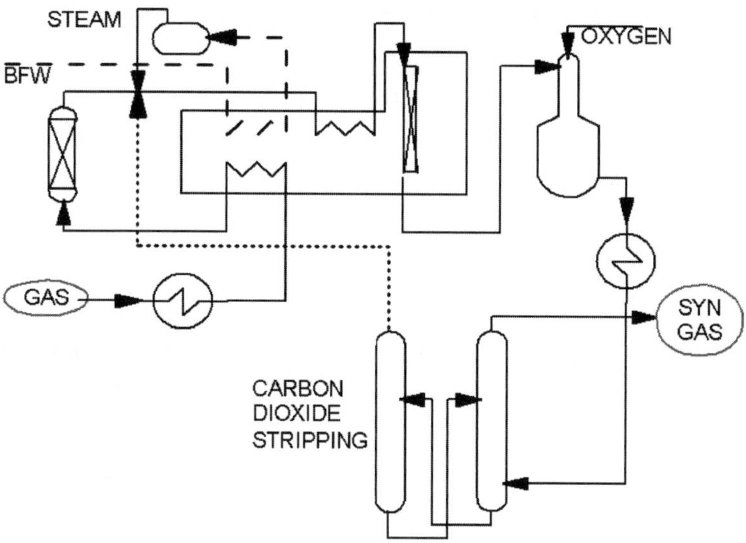

Fig. 7–4. High carbon monoxide gas

The current interest in ameliorating greenhouse emissions, principally carbon dioxide, has already been mentioned. However, it is often rather naively expressed that the reverse WGS can be used to eliminate carbon dioxide and produce more valuable product.

A quick analysis of the thermodynamics of the conversion of carbon dioxide reveals the futility of this approach. Processes that consume carbon dioxide require substantial energy input. Greenhouse gas amelioration will only occur if this heat input is from a renewable source. This is because using fossil fuel energy will result in more carbon dioxide emission (from the fuel) than can be consumed by the carbon dioxide reactions of the WGS.

Downstream Processes

There are large numbers of processes that utilize synthesis gas. Individual processes require adjusting the stoichiometric ratio by WGS or extraction of components. Table 7–1 summarizes the major processes and gives the ideal stoichiometric ratios.

Table 7–1. Major processes requiring synthesis-gas

Product	Required Stoichiometric Ratio (a)	Max CO_2 (mol%)	Comment
Methanol	2:1	Not an issue (b)	Main problem from steam reformed gas is excess hydrogen
Ammonia	Maximum	nil	Hydrogen required as intermediate
Fischer-Tropsch: CO catalysts	2:1		Some CO_2 can be tolerated
Fischer-Tropsch: Fe catalysts	1.7/1	<5%	Extra hydrogen made by internal WGS
Oxo Process	1:1	nil	Hydrogen required in subsequent alcohol production step
Directly reduced iron	3:1	<10%	Water and CO_2 kept to a minimum
Acetic Acid	CO	Nil	Carbon monoxide

Notes: (a) stoichiometric ratio (mole% hydrogen/mole% carbon monoxide); (b) carbon dioxide improves process kinetics

Ammonia synthesis

The objective is to obtain an ammonia synthesis gas of composition 3:1 hydrogen: nitrogen.

$$3H_2 + N_2 = 2NH_3 \tag{7.8}$$

Natural gas provides the hydrogen, and air the nitrogen. The conventional approach uses a two-stage process that comprises a steam reformer and a partial oxidation unit using air as the oxidant. In the first stage, steam reforming is conducted at a pressure of about 30 bar at 800°C or higher. The natural gas is only partially reformed, and sufficient methane is maintained in the stream for the second stage.

In the second stage, the partially reformed gas is mixed with air and passed over a catalyst. This promotes the oxidation of the residual methane into synthesis gas at about 1,000°C. Typical gas compositions are given in table 7–2.

Table 7–2. Typical gas compositions in ammonia synthesis

	Gas Feed to Reformer	Exit Reformer	Exit Partial Oxidation
Pressure (bar)	35	30	29
Temperature (°C)	525	790	971
Methane (mol%)	91.9	9.4	0.2
Ethane (mol%)	2.9		
Propane (mol%)	0.6		
Butanes (mol%)	0.2		
Pentanes (mol%)	0.1		
Hexanes + (mol%)	0.1		
Carbon dioxide (mol%)	0.3	11.6	8.8
Nitrogen (mol%)	1	0.5	22.1
Carbon monoxide (mol%)		8.3	11.5
Hydrogen (mol%)	2.9 (a)	70.2	57.1
Argon (mol%)			0.3

Note (a) Small amount of hydrogen added to feed to ensure catalyst in the top of the reformer remains reduced. Data from ICI Ltd. *Catalyst Handbook*, M. V. Twigg, editor, London: Wolfe Scientific Books, 1970; and L. Lloyd, D. E. Ridler, and M. V. Twigg. *Catalyst Handbook*, second edition, M. V. Twigg, editor, London: Wolfe Publishing Ltd., 1989.

Downstream from the synthesis gas section, the WGS reactor shifts the carbon monoxide to carbon dioxide. This is subsequently stripped, and residual carbon oxides are converted to methane before ammonia synthesis. The detail and economics of ammonia synthesis are discussed further in a later chapter.

Methanol synthesis

Since only hydrogen and carbon oxides are required, the objective is to convert as much of the methane as possible. In steam reformer operations, this requires the pressure in the reformer to be low, and the temperature to be as high as economically feasible. Excess hydrogen is extracted from the methanol synthesis loop and used as fuel for the reformer.

Modern plants use either a hybrid system or small partial oxidation unit (oxygen secondary reforming). These are used in order to minimize methane slip and avoid excess hydrogen in the methanol synthesis.

The details and economics of methanol synthesis are discussed further in a later chapter.

Oxo synthesis gas

The oxo process is the addition of hydrogen and carbon monoxide to an olefin to produce an aldehyde. The oxo reaction is used to produce the plasticizer 2-EH (2-ethylhexanol) and biodegradable detergents. The reaction is given in equation 7.9, where R is an alkyl group.

$$R\text{–}CH{=}CH_2 + H_2/CO = R\text{-}CH_2\text{-}CH_2\text{-}CHO \qquad (7.9)$$

The requirement for a synthesis gas with a stoichiometric ratio of unity makes partial oxidation a useful procedure.

Oxo synthesis gas can also be made using a steam reformer by adding carbon dioxide to the gas feed and minimizing the steam/methane ratio. This promotes the process:

$$CH_4 + CO_2 = 2CO + 2H_2 \qquad (7.10)$$

To achieve the required synthesis gas, relatively low pressures and temperatures are used. Typical gas compositions are given in table 7–3.

Table 7–3. Typical steam reformer gas compositions in the Oxo synthesis

	Gas Feed to Reformer	Exit Reformer
Pressure (bar)	16.5	13.5
Temperature (°C)	540	870
Methane (mol%)	62.2	2.9
Ethane (mol%)	1.8	
Propane (mol%)	0.3	
Butanes (mol%)	0.3	
Pentanes (mol%)	0.1	
Carbon dioxide (mol%)	29.6	10.8
Nitrogen (mol%)	0.6	0.2
Carbon monoxide (mol%)	2.1(a)	21.9
Hydrogen (mol%)	3.0(a)	64.2

Note (a) recycle gas to maintain the catalyst in a reduced state (data based on Ridler and Twigg)

Reverse WGS can be used to convert the carbon dioxide into carbon monoxide, essentially extracting the carbon dioxide and recycling it to the reformer. One of the problems with the production of synthesis, as with a low stoichiometric ratio, is the tendency to form carbon by the Boudart reaction (eq. 7.11).

$$2CO = CO_2 + C \qquad (7.11)$$

In addition, there is the decomposition of methane to its elements (eq. 7.12):

$$CH_4 = C + 2H_2 \qquad (7.12)$$

These problems have been addressed by workers at Haldor-Topsoe A/S in the development of the sulfur-passivated steam reforming (SPARG) process and in the development of their proprietary advanced reforming process.[6] These processes use a sulfur-passivated catalyst to produce low stoichiometric synthesis gas from high carbon dioxide feed. This is accomplished through high-temperature reforming.

Direct iron ore reduction—reducing gas

Reducing gas has the duty to reduce iron ore to metal. A high concentration of hydrogen is required. This is produced by steam reforming at high temperatures and low pressures (below 10 bar), with a low steam/methane ratio. Typical reformer inlet and exit compositions are shown in table 7–4.

Table 7–4. Steam reformer gas compositions for reducing gas

	Gas Feed to Reformer	Exit Reformer
Pressure (bar)	10.7	7
Temperature (°C)	510	830
Methane (mol%)	85.5	4
Ethane (mol%)	9.1	
Propane (mol%)	0.6	
Carbon dioxide (mol%)	3.7	7
Nitrogen (mol%)	1.1	1
Carbon monoxide (mol%)		16
Hydrogen (mol%)		72

Data based on Ridler and Twigg

Hybrid Systems

An obvious objective of technology development is to combine the benefits of steam reforming (high stoichiometric ratio, low capital cost) with those of partial oxidation. Benefits of the latter include high-pressure operation and low methane slip. The approach is to use the heat of partial oxidation to drive the steam reforming process in a single-reactor system. Another benefit of the integration is to improve the match of the downstream stoichiometric ratio of the product synthesis gas with that required for the downstream process. This will minimize and even possibly avoid the WGS or extraction processes. There have been many attempts to do this, with varying degrees of success. Two such processes are autothermal reforming and gas-heated reforming.

Autothermal reforming

In this process natural gas, oxygen, and steam are passed to a combustion lance in the top of the reactor. The gas is only partially converted into synthesis gas but is heated to the reforming temperature. The hot gases then pass over a reforming catalyst in the base of the reactor. After leaving the reactor, the synthesis gas is cooled. In the process, steam is raised to produce process and export steam (fig. 7–5).[7]

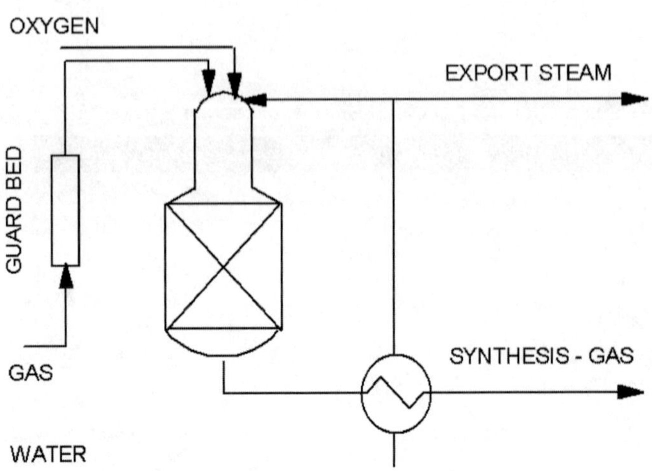

Fig. 7–5. Autothermal reforming

Gas-heated reformer (GHR)

The gas-heated reformer works in the opposite manner of the autothermal reactor. After saturation, the gas is passed to a reformer, where a catalyst partially reforms the gas into synthesis gas. The gases are then passed to a secondary reformer, where combustion with oxygen over a catalyst takes place. The secondary reformer converts all of the residual methane into synthesis gas. The hot gases produced in the secondary reformer are then passed back to the primary reformer, where they provide the process heat for the primary reforming. This is illustrated in figure 7–6.

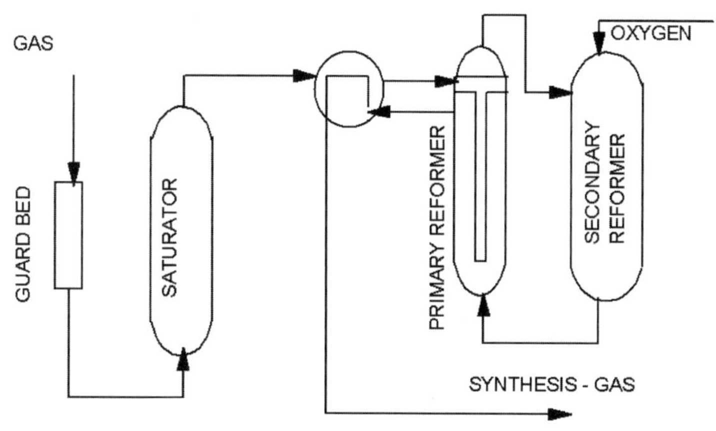

Fig. 7–6. Gas-heated reformer

This approach was developed by ICI Ltd. as part of its ICI LCA ammonia process.[8] It was then further developed for the production of methanol by BHP Ltd. and is proposed as a method for producing synthesis gas for the FT process.[9] Significant improvements in thermal efficiency compared to conventional technology are claimed.

New Developments

Ion-exchange membrane processes

The major cost in the partial oxidation route is the cost of oxygen. The ceramic ion-exchange membrane technology, which is currently under development by several groups, aims to eliminate the oxygen requirement. In this process, a special ion-exchange ceramic membrane allows oxygen from the air to permeate through a ceramic membrane and oxidize natural gas on the other side.

The ceramic is a composite mixture, often of doped zirconia or yttria, which at high temperature permits the transfer of oxygen ions in one direction and electrons in the other. The operation of the membrane is illustrated in figure 7–7.

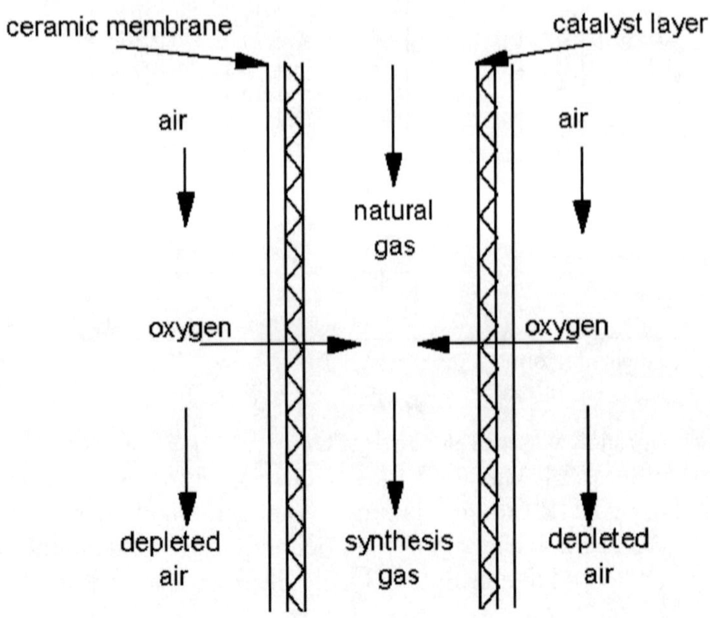

Fig. 7–7. Ion-exchange membrane for synthesis gas

In operation, the system would comprise many ceramic tubes through which the natural gas would pass, broadly similar to a steam reformer. In order to promote combustion, the hydrocarbon side is coated with a porous oxidation catalyst. The resulting combustion of the gas results in typical temperatures in the range 1,000°C to 1,300°C.

The ion exchange membrane system offers the advantage of eliminating the oxygen plant, but at this stage in its development, several problems are apparent:

- It is not clear how the ceramic will stand up to the rigors of high-temperature and high-pressure operation. To date, demonstration plant efforts have concentrated on low-pressure operation to control large differences in temperature both axially and laterally.
- The temperature of operation is lower than that of a normal partial oxidation unit. This results in excessive carbon dioxide, which has to be removed and recycled for some uses.
- Combustion rates are determined by the rate at which oxygen and counter ions can move across the ceramic and catalyst layers. This results in the requirement of a large surface area for effective combustion (i.e., a large number of tubes).

One prospective use of an ion-exchange membrane is as a partial oxidation reactor placed after a steam reformer. When a conventional system is used, this is known as oxygen secondary reforming. This process produces a synthesis gas that is rich in carbon monoxide, which is useful for petrochemical operations. An ion-exchange version of such a system is described by Nataraj and others.[10] A membrane separator is placed after the reformer and prior to the removal of carbon dioxide, which is recycled. The authors describe three cases. The first case has no reformer (i.e., a membrane-only system), the second has an adiabatic reformer, and the third uses a high-temperature (tubular) reformer. Table 7–5 shows the salient results converted to metric units.

Table 7–5. Synthesis-gas composition using a membrane system with differing reformers

		No Reformer	Adiabatic Reformer	High Temp. Reformer
Syn gas	GJ/h	2656.8	2663.8	2668.2
CH_4	Mol%	0.5	0.5	0.5
CO	Mol%	16.5	15.3	14.7
CO_2	Mol%	12.4	13.3	13.7
H_2	Mol%	70.1	70.4	70.6
CO_2 removed	kgMol/h	387.8	159.2	None
Oxygen	T/d	1691.9	1350.8	1111.3
Gas used	GJ/h	3122	2929.9	2794.9
Fuel	GJ/h	288	444.2	192
Export steam	GJ/h	3.4	2.6	-5.4
Power demand	GJ/h	27	21.2	16.4
Power fuel (a)	GJ/h	60	47.2	36.4
Total input	GJ/h	3466.6	3418.7	3028.8
Process efficiency	%	76.6	77.9	88.1

Note (a) Fuel required to supply the power at an efficiency of 45% (gas-turbine combined-cycle).

The cases illustrate the production of a synthesis gas of similar quantity and quality and then estimate the overall energy required. The table illustrates that placing the membrane separator after a reformer improves the overall efficiency. A high temperature reformer/membrane separator combination gives very high efficiencies.

Note that this combination achieves the required synthesis gas composition without the need for any carbon dioxide extraction and recycling. This eliminates the energy demand in driving the highly endothermic carbon dioxide conversion processes.

Partial oxidation with air

If a downstream process can be adapted or designed to operate in the presence of nitrogen (as in the case for ammonia synthesis), then partial oxidation with air may be feasible. The main interest in this is the FT process, where *some* systems are tolerant to high levels of nitrogen. This has led some groups to research partial oxidation of natural gas with air.

The presence of high levels of nitrogen results in a much cooler process, but in order to achieve useful yields of synthesis gas, a catalytic system is required. Using a nickel catalyst, Jess and Hedden have obtained a synthesis gas of the following composition:[11]

Methane	2.4%
Carbon monoxide	15%
Hydrogen	30%
Carbon dioxide	2.6%
Water	5.2%
Nitrogen	44.8%

Ignoring the nitrogen, the gas composition is very similar to that produced by a steam reformer operating under low severity conditions. Costs are saved in the partial oxidation reactor by elimination of the air separation plant. However, the effect of the nitrogen and other impurities significantly increases the separation costs in the downstream process operations.

Economics of Synthesis Gas Production from Natural Gas

Steam reforming

The basis for this analysis is a steam reformer plant that produces a gas of stoichiometric ratio 3:1, with minimal carbon dioxide. The plant is a stand-alone facility equipped with WGS units and hydrogen extraction units so as to be able to produce gas of varying SR. The plant is a large-scale operation annually producing 644 kt of gas.

Different organizations express synthesis gas usage in quite varied ways. Some equivalents for the previous are given in table 7–6.

Table 7–6. Synthesis gas equivalent measures

kt/a	643.88
kt/d	1.89
PJ/a	21.62
MMcm/d	5.28
MMcf/d	186.68
Ncm/h	219828

The fixed-variable cost of production (with gas as the variable cost) of synthesis gas is illustrated in figure 7–8. Pertinent statistics for the plant are given in table 7–7.

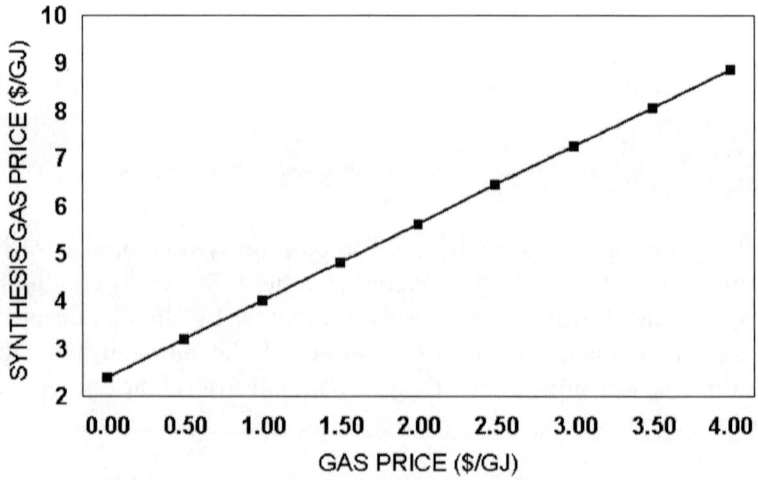

Fig. 7–8. Synthesis gas costs by steam reforming

Table 7–7. Statistics for synthesis-gas production using steam reforming

		$CO/3H_2$
Production	kt/a	643.9
Capital cost	$MM	219.3
Construction period	years	3
Plant life	years	15
Return on capital (10% DCF)	%/a	16.34%
Non gas operating costs	$MM/a	53.96
Gas usage	PJ/a	35.02
By-product credits (electricity)	$MM/a	2.08

In the production of synthesis gas using a steam reformer, there is an excess of steam produced. The handling and disposal of this steam can profoundly impact the overall economics. In this concept, about 1 PJ/y of excess steam is generated. This can be used to produce power by a condensing generator (25% efficiency), with a value of the electricity exported of 3¢/kWh. It gives an annual by-product credit of more than $2 million.

Partial oxidation

In contrast to a steam reformer, which is a large plant, a partial oxidation plant (the gasifier) is quite compact. This results in much less capital for the basic process plant, such as illustrated in figure 7–2. In fact, if heat recovery is not a necessity, further savings can be made to the basic process by using a quench system as opposed to a waste heat boiler.

Most of the capital is associated with the oxygen plant required to operate the partial oxidation. For normal operation, more oxygen in weight terms is required than natural gas. The oxygen plant is relatively large, and the production of oxygen dominates the process economics.

An advantage of the partial oxidation system is that it will produce synthesis gas low in carbon dioxide and of the correct stoichiometric ratio for some immediate uses. This can eliminate the need for WGS units and carbon dioxide extraction systems.

Thus the cost for the production of a raw synthesis gas of SR = 1.75 is lower than that for steam reforming, which requires extensive WGS and carbon dioxide extraction. The fixed variable relationship is illustrated in figure 7–9 for two scales of operation that are detailed in the table 7–8.

For this concept estimate, all excess steam is assumed to be consumed in powering the air separation plant. This can be accomplished with steam or electric drives, or a combination of both. As a consequence, no by-product electricity results.

A comparison can be made between figures 7–8 and 7–9 relative to these process economics. If the required synthesis gas can be produced from partial oxidation, it is apparent that this is a lower cost route. This is true even for a small-scale process.

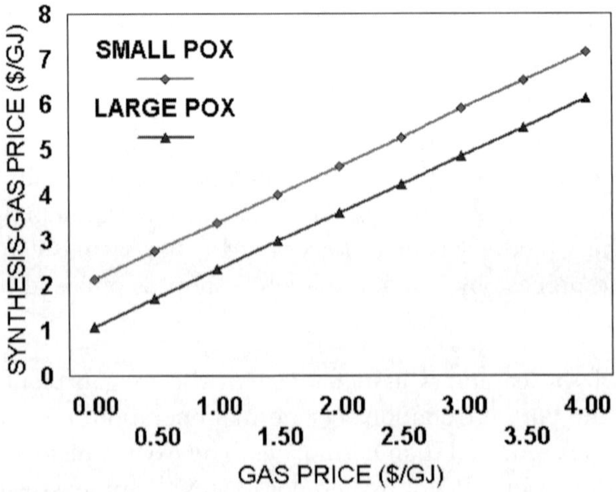

Fig. 7–9. Synthesis gas costs by partial oxidation

Elimination of process related to WGS, and in particular carbon dioxide removal, will increase the overall efficiency of the process. This is typically in the range 78% to about 83% for the simple partial oxidation systems, compared to 60% to 65% for steam reformers equipped with WGS and carbon dioxide removal. However, should the stoichiometric ratio of the synthesis gas require adjustment, by WGS and/or carbon dioxide removal, the benefits of the partial oxidation route are lost.

Table 7–8. Statistics for synthesis gas production using partial oxidation

		Small	Large
Production	kt/a	161.4	1078.0
Capital cost	$MM	31.6	108.5
Construction period	years	3	3
Plant life	years	15	15
Return on capital (10% DCF)	%/a	16.34%	16.34%
Non gas operating costs	$MM/a	6.89	23.69
Gas usage	PJ/a	4.07	27.19

Alternative approaches for methanol production. The following data were produced during the mid-1980s for a world-scale (2,000 t/d) methanol plant by workers at BHP. Methanol is best produced from a synthesis gas of stoichiometric ratio about 2. There are no restrictions regarding the carbon dioxide content. Table 7–9 compares the relative capital costs for producing a suitable synthesis gas from steam reforming, partial oxidation, or autothermal reforming.

Table 7–9. Relative cost of producing synthesis gas by various routes

Unit Operation	Reformer	Partial Oxidation	Autothermal
Reactor	100	54	59
Air separation	0	46	41
Overall relative cost	100	104	100

In this case, partial oxidation was found to be more capital intensive than conventional steam reforming, but it was similar to autothermal reforming in cost. A similar result has been reported by KTI for the production of synthesis gas for the subsequent manufacture of carbon monoxide.[12]

However, examining the detail of the synthesis gas composition produced by the various alternative processes reveals a further advantage for the autothermal route. The hydrogen content is a closer match to that required to minimize recycling compression costs in the methanol synthesis. Observations such as this have led to growth in the hybrid processes for producing specific synthesis gases for specific end uses.

Carbon Monoxide

Carbon monoxide is required for the production of a variety of chemicals, including acetic acid and phosgene (for the synthesis of isocyanates). Carbon monoxide is produced from synthesis gas by either low-temperature (cryogenic) separation or by reversible absorption in a copper solution. Membrane permeation and PSA can also be used to extract carbon monoxide. As a result of the production of carbon monoxide, the hydrogen component of the synthesis gas is separated as a by-product. The block flow for producing carbon monoxide by cryogenic separation is illustrated in figure 7–10.

The synthesis gas is passed to a carbon dioxide absorber, where the dioxide content is reduced to about 50 ppm. This is followed by drying and final carbon dioxide removal using a molecular sieve. The dry gas is then cooled to about –180°C, and at a pressure of 40 bar, the carbon monoxide and methane are condensed. Hydrogen remains in the gaseous state and is removed. The liquids are passed to a distillation column and depressurized to about 2.5 bar to separate the carbon monoxide from the methane.

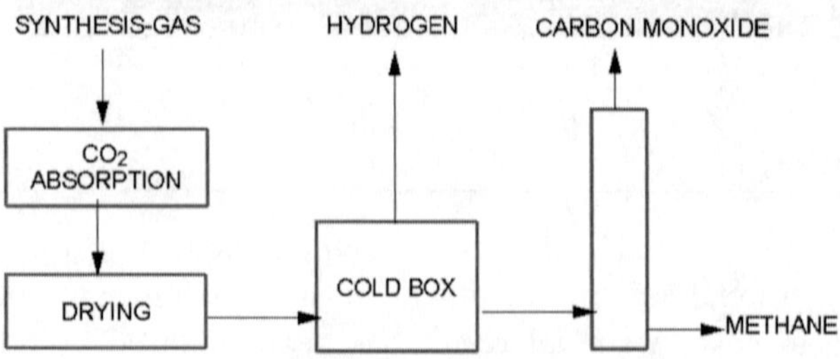

Fig. 7–10. Carbon monoxide production

KTI has reported a quantitative analysis of the alternative methods for the production of carbon monoxide from a wide variety of feedstocks.[13] Its analysis also included a comparison of the alternative separation processes. The salient results given in table 7–10 show the benefits of the copper absorption process relative to the alternatives.

Table 7–10. Comparison of CO separation and purification processes (3000Ncm/h, 10 bar)

Process	COSORB (a)	Cryogenic	Cryogenic + methane wash	Membrane permeation	PSA
Permeation					
Typical Pressure (bar)	2–30	22–30	20–50	12–50	8–30
Steam use	Stripper reboiler	Dryer regen.	Dryer regen.		
CO recovery (%)	99	96.8	98	90.7	90.5
CO purity (%)	97.9	96.8	99.8	97	96.8
Impurities	CO_2	H_2, CH_4	H_2, CH_4	CH_4	CH_4
Relative Opex (b)	1	1.072	1.086	1.081	1.294
Relative Capex (c)	1.041	1	1.025	1	1.1

Notes (a) Uses cuprous aluminium chloride ($CuAlCl_4$) in toluene to form a loose complex with CO, the absorbent is regenerated by heating; (b) Utilities, (c) Annual capital charges (25%) plus labor and maintenance.

Hydrogen

After production by any of the above processes, the synthesis gas can be converted entirely into hydrogen. Hydrogen is used in a wide variety of chemical and refining operations and is of current interest as an alternative fuel in the so-called Hydrogen Economy. Figure 7–11 illustrates a common approach to the production of hydrogen from synthesis gas.

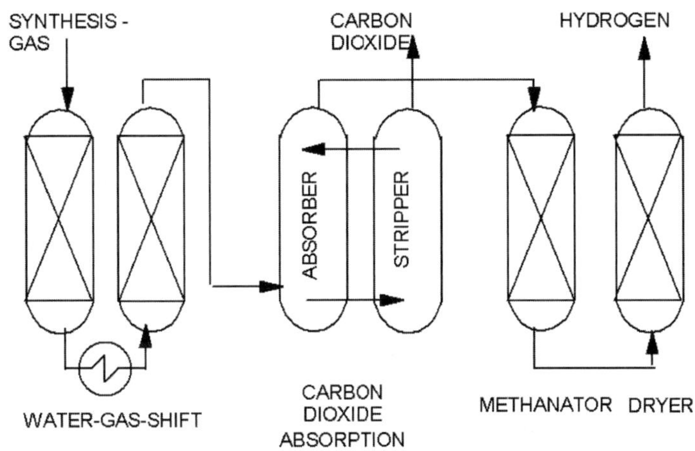

Fig. 7–11. Hydrogen production

The objective of the process operations is to remove all of the carbon oxides present (which usually interfere with the downstream process using hydrogen) and to maximize the yield of hydrogen. The first operation is that of WGS, which converts the carbon monoxide into carbon dioxide and produces extra hydrogen by the following reaction:

$$CO + H_2O = CO_2 + H_2 \qquad (7.13)$$

This process is usually conducted in two stages (known as HT and LT shift) over different catalysts (see previous discussion). The next stage is the removal of carbon dioxide, usually by an absorber/stripper system. This uses very similar technology to an acid gas plant for the removal of hydrogen sulfide and carbon dioxide from natural gas.

Methanol can be added to the shift converter in order to increase yields.[14] This is accomplished by the following reaction:

$$CH_3OH + H_2O = CO_2 + 3H_2 \tag{7.14}$$

The gas still contains traces (in parts per million) of carbon oxides. These are removed in the methanator that converts them into methane, which is inert to most of the downstream uses of hydrogen. This sacrifices some hydrogen, but the loss is small (< 1%).

$$CO + 3H_2 = CH_4 + H_2O \tag{7.15}$$

$$CO_2 + 4H_2 = CH_4 + 2H_2 \tag{7.16}$$

In the final step, the gas is then dried to remove water.

In addition to the total conversion of synthesis gas into hydrogen, excess hydrogen can be removed from a synthesis gas of high stoichiometric ratio by PSA.[15] This can also be accomplished by using membrane separators.[16] These can be adapted to produce only hydrogen by extracting as much hydrogen as possible and returning the carbon monoxide synthesis gas to fuel the reformer.

If methane contamination of hydrogen is a problem, then very pure hydrogen can be produced by methanol cracking. In this route, methanol and steam are passed over a catalyst at 250°C to 300°C. This splits the methanol into hydrogen and carbon oxides by the reaction given in equation 7.14 and by the following:

$$CH_3OH + H_2O = CO + 3H_2 \tag{7.17}$$

Cryogenic fractionation or absorption removes carbon oxides. Hydrogen produced by this method is somewhat more expensive than hydrogen produced directly from synthesis gas. However, the purity of the product is attractive to some end users who require particularly high purity hydrogen (such as liquid hydrogen rocket fuel).

Hydrogen production costs

Hydrogen production is best accomplished by the steam reformer method of producing synthesis gas, since this process gives a high hydrogen stoichiometric ratio gas and does not require the provision of an expensive oxygen plant.[17]

The fixed variable relationship for four scales of production (in tonnes per day) is shown in figure 7–12, with the pertinent statistics given in table 7–11.

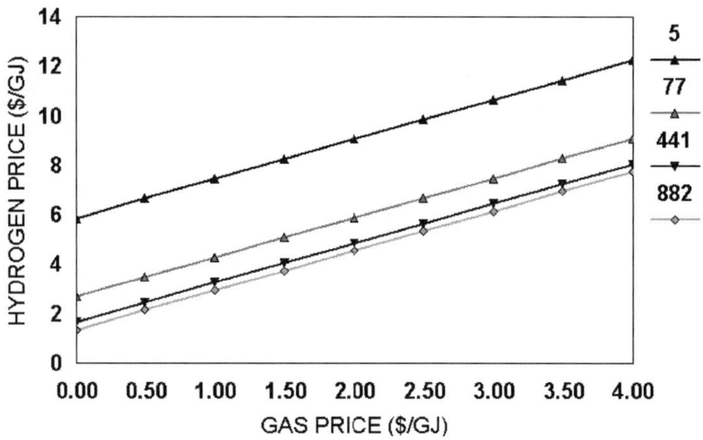

Fig. 7–12. Hydrogen production costs

Table 7–11. Statistics for hydrogen production

		H_2 micro	H_2 small	H_2	H_2 large
Production	kt/a	1.76	26.20	150	300
Capital cost	$MM	6.66	55.65	219.34	378.20
Construction period	years	3	3	3	3
Plant life	years	15	15	15	15
Return on capital (10% DCF)	%/a	16.34%	16.34%	16.34%	16.34%
Non gas operating costs	$MM/a	1.49	12.68	50.91	88.62
Gas usage	PJ/a	0.40	5.94	34.00	68.00
By product credits (electricity)	$MM/a	0.02	2.66	15.25	30.50

Again there are alternative ways of expressing the production of hydrogen. The equivalent table for the preceding cases is given in table 7–12.

Table 7–12. Statistics for hydrogen production

	H_2 micro	H_2 small	Base H_2	H_2 large
kt/a	1.76	26.2	150	300
t/d	5.2	77	441.2	882.4
PJ/a	0.25	3.71	21.27	42.54
MW	8.5	126.14	723.18	1446.36
MMcm/d	0.06	0.9	5.18	10.36
MMcf/d	2.15	32	183.23	366.47
Ncm/h	2530	37683	215772	431545

The base case hydrogen plant is the same process plant used for the production of synthesis gas as previously discussed. In this case, the WGS plant and carbon oxide removal system serve to remove all of the carbon present. This will produce 150 kt/y (441 t/d) of hydrogen product. By-product credits are generated from electricity from by-product steam (6.7 PJ/y). This has been converted into electricity on the same basis as previously described. This base case is scaled to 300 kt/y (882 t/d) hydrogen to illustrate the cost for the mass production of hydrogen for a hydrogen economy. The scaling factor used is 0.786.[18]

Using the same scaling factor, the base case has been downsized to an output of 26 kt/y (77 t/d, 32 MMscfd). This is typical of many smaller manufacturing requirements for hydrogen, such as the provision of extra hydrogen for refinery operations. A final case of 5 t/d is used to illustrate very small (micro) plants that may be required for small-scale local hydrogen production.

Although steam reforming seems the preferred method for producing hydrogen from natural gas, there are occasions when partial oxidation can offer an economic alternative. This occurs when there is a large demand for the by-product gases from the air separation plant. In this regard BOC Ltd has described the production of 32 MMscfd hydrogen for refinery use, using partial oxidation of natural gas as opposed to steam reforming.[19] This required the use of a large air separation plant that produced additional speciality gases for local uses.

Stoll and Von Linde have reported a comparative cost analysis for the production of hydrogen.[20] They compared production costs using electrolysis and methanol cracking with steam reforming over a range of scales of production. They showed that steam reforming was the cheapest production method for volumes greater than about 200 Nm^3/h. Salient data are illustrated in figure 7–13. The Stoll and Von Linde data relate particularly to very small plants. The data for the microhydrogen plant discussed previously are placed on their data set for comparison.

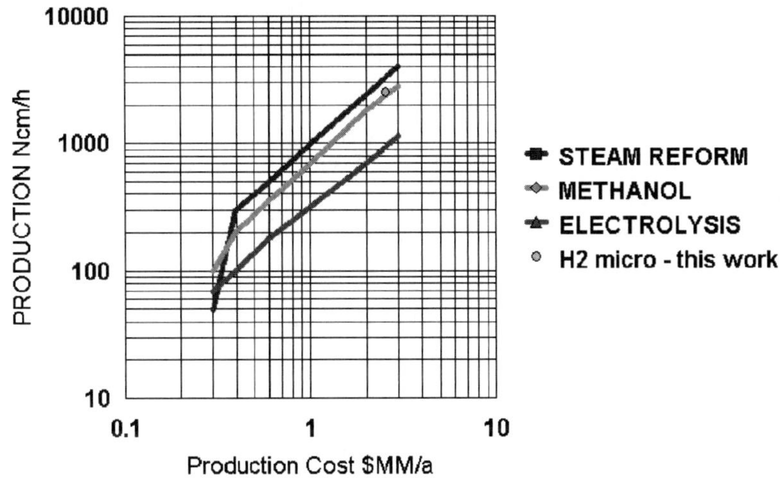

Fig. 7–13. Comparative hydrogen production costs

Greenhouse Gas Implications

Due to the current concern with greenhouse gas emissions, particularly carbon dioxide, it is relevant to estimate the emissions resulting from the production of synthesis gas and hydrogen. Greenhouse gas emissions result from several sources:

- **Direct emission of carbon dioxide to atmosphere.** This occurs when the objective is to produce hydrogen gas or synthesis gas of a high stoichiometric ratio. In this case, the natural gas is used to produce hydrogen. The carbon in the gas is converted to carbon dioxide and then is extracted and expelled to atmosphere.

- **Fuel consumed in driving the reactions.** For steam reforming, this is mainly the fuel consumed in the reformer.
- **Fuel consumed in driving ancillary equipment.** For partial oxidation, the main energy used is in the power consumption associated with the air separation plant. If power (or oxygen) is purchased from another party, then greenhouse costs may be reflected in the greenhouse emissions of the other party.
- **Emissions from the use of utilities or credits for exported by-products.** This principally involves the greenhouse cost for the production of imported steam and electricity or credits for steam or power exported and used in other plants.
- **Fugitive emissions of gases.** Of particular concern is the fugitive emission of natural gas because methane (the main component) is 21 times more greenhouse intensive than carbon dioxide.
- **Emissions of combustion gases containing nitrous oxide.** When fuels are burned, the high temperature results in the reaction of oxygen and nitrogen (in the air or in the fuel) to form nitrogen oxides. A small portion of the nitrogen oxides is comprised of nitrous oxide, which is 310 times more greenhouse intensive than carbon dioxide. Because substantial volumes of exhaust gases are emitted, the emission of nitrous oxide in greenhouse terms can be significant.

From the above it is evident that each facility would have to be analyzed on an individual basis (called a *cradle-to-grave analysis*). In broad terms, the first two areas of emission result in the largest greenhouse impact. Thus some estimates of greenhouse impacts based on the overall energy efficiency can be made.

Estimates can be made from the results of a synthesis gas in which no carbon dioxide is extracted. Alternately, estimates can be made from hydrogen in which all of the carbon in the feed is extracted and discharged as carbon dioxide. The estimates are for the production of 634 kt/y of synthesis gas and 150 kt/y hydrogen. The economic cases are given in the previous sections, with gas at $1/GJ (as representative of large-scale production of hydrogen for a hydrogen economy).

The cost of discharge of carbon dioxide could be considered to be US$10/t, levied either as a carbon emissions tax or a requirement to purchase offsetting credits. In this case, the cost of production of synthesis gas rises by about 10%, and that for hydrogen by 20%.

The sensitivity of the unit production cost to the cost of carbon dioxide discharged is illustrated in figure 7–14. The graph illustrates the dramatic impact of greenhouse amelioration strategies based on carbon taxes on the production cost of hydrogen.

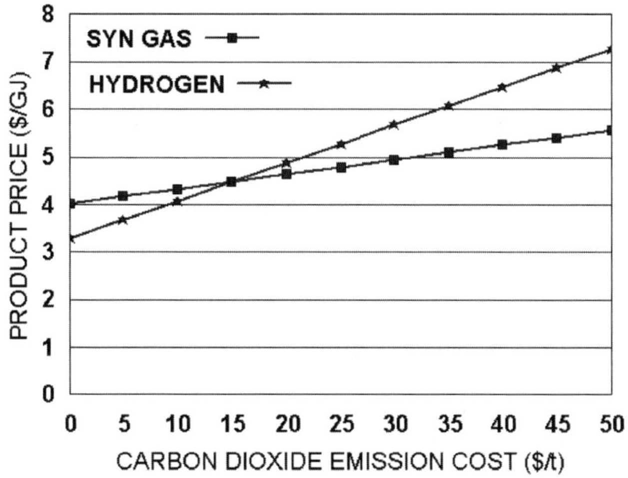

Fig. 7–14. Impact of greenhouse emission cost

References

1. Gunardson, H. H., and J. M. Abrado. "Produce CO-Rich Synthesis Gas," *Hydrocarbon Processing*, April 1999, p. 87.

2. ICI Ltd. *Catalyst Handbook*, M. V. Twigg, editor, London: Wolfe Scientific Books, 1970; and L. Lloyd, D. E. Ridler, and M. V. Twigg, *Catalyst Handbook*, second edition, M. V. Twigg, editor, London: Wolfe Publishing Ltd., 1989.

3. Montgomery, C. W., E. B. Weinberger, and D. S. Hoffman. "Thermodynamics and Stoichiometry of Synthesis Gas Production," *Industrial & Engineering Chemistry*, 40(4), 1948, p. 601.

4. Twigg, M. V., editor. *Catalyst Handbook*, second edition, 1989, p. 283.

5. UOP LLC. "Gas Processes 2002," *Hydrocarbon Processing,* May 2002, p. 70.

6. Dibbern, H., et al. "Make Low H_2/CO Syngas Using Sulfur Passivated Reforming," *Hydrocarbon Processing*, January 1986, p. 71; Haldor Topsoe A/S. "Gas Processes 2002," *Hydrocarbon Processing,* May 2002, p. 93, and references therein.

7. Howe-Baker Engineers Ltd. "Gas Processes 2002," *Hydrocarbon Processing,* May 2002, p. 103.

8. ICI Katalco. "Petrochemical Processes 95," *Hydrocarbon Processing,* March 1995, p. 95.

9. Abbott, J., and B. Crewdson. "Gas Heated Reforming Improves Fischer-Tropsch Process," *Oil & Gas Journal*, April 22, 2002, p. 64.

10. Nataraj, S., R. B. Moore, and S. L. Lee. "Production of Synthesis Gas by Mixed Conducting Membranes," U.S. Patent 6,048,472, April 11, 2000 to Air Products and Chemicals Inc.

11. Jess, A., and K. Hedden. "Production of Synthesis Gas by Catalytic Partial Oxidation of Methane with Air," *Oil Gas—European Magazine,* April 1994, p. 23.

12. "Carbon Monoxide Production Technologies," *KTI Newsletter*, Winter 1987.

13. Ibid.

14. Dybkjaer, I., "What Are the Options for Hydrogen Plant Revamps?" *Hydrocarbon Processing*, February 2005, p. 63.

15. Watson, A. W. "Use Pressure Swing Adsorption for Lowest Cost Hydrogen," *Hydrocarbon Processing*, March 1983, p. 91.

16. Shaver, K. G., G. L. Poffenbarger, and D. R. Grotewold. "Membranes Recover Hydrogen," *Hydrocarbon Processing*, June 1991, p. 77.

17. Seddon, D. "What Price Hydrogen?" Chemica 2004, Australian Technology Centre, Sydney, September 27–30, 2004.

18. From data given by Foster Wheeler in "Refining Processes 2002," *Hydrocarbon Processing,* p. 123.

19. Menon, R., et al. "Industrial Gas Partnership Enables Upgrade at BP Refinery," *Oil & Gas Journal*, March 19, 2001, p. 70.

20. Stoll, R. E., and F. von Linde. "Hydrogen—What Are the Costs?" *Hydrocarbon Processing*, December 2000, p. 42.

8

METHANOL AND DERIVATIVES

This chapter deals with methanol and its derivatives—dimethyl ether, formaldehyde, and acetic acid. Methanol is an important chemical, with the world annual production being about 30MMt. Natural gas is the favored feedstock for the production of chemical methanol, and it is produced in large plants where sufficient volume of gas is available.

From the standpoint of the Gas Age, methanol is potentially an important gas-derived fuel in its own right. It is also an intermediate to other alternative fuels, such as dimethyl ether (DME). To produce alternative fuels, there would need to be a significant change toward mass production of methanol on a much larger scale than currently exists.

Methanol

Methanol production technology

Methanol is produced from synthesis gas by catalysts with the overall reaction stoichiometry:

$$CO + 2H_2 = CH_3OH \tag{8.1}$$

Carbon dioxide is also converted in the reaction and is not removed from the synthesis gas:

$$CO_2 + 3H_2 = CH_3OH + H_2O \tag{8.2}$$

Gas free of carbon dioxide is converted more slowly than gases containing the dioxide. Thus it appears the preferred reaction route is via carbon dioxide with carbon monoxide being converted into carbon dioxide by the WGS reaction in situ (see chapter 7).

The formation of methanol is exothermic, and as a consequence, low temperatures favor the reaction. The main issue is removing the reaction heat, and there are several proprietary engineering approaches to this problem. This is extensively discussed in texts specifically concerned with methanol synthesis.[1]

Figure 8–1 shows the principal unit operations for the production of methanol from natural gas. Traditionally, the production of synthesis gas is by a steam reformer. After the reformer, the synthesis gas requires compression via the synthesis gas compressor. Older processes operate at very high pressures (typically 200 bar), but more modern processes operate at much lower pressures. The low-pressure synthesis route dominates the present technology and can produce methanol at pressures of about 50 bar. However, it more usually operates at pressures near 100 bar. This facilitates higher pass conversion or lower operational temperature. The synthesis gas therefore requires compression from about 30 bar to about 100 bar.

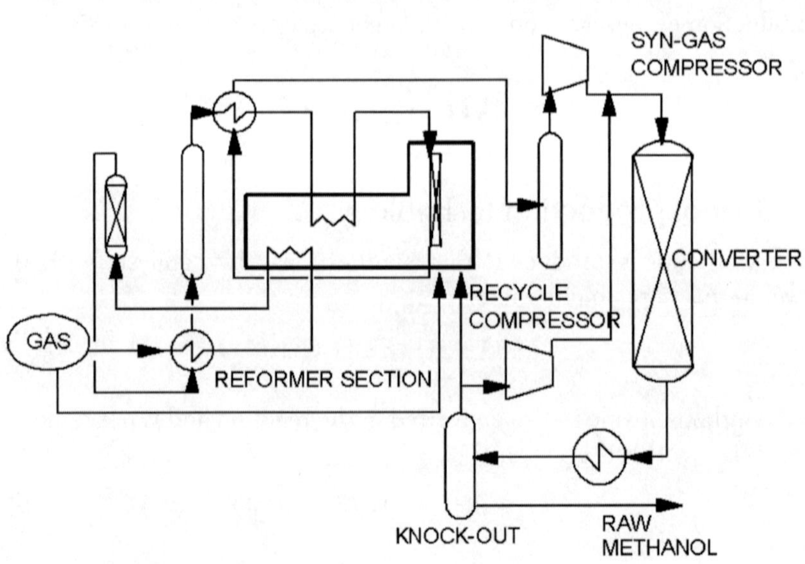

Fig. 8–1. Methanol synthesis

The synthesis gas is passed to the converter, with the conversion usually taking place at about 300°C. The pass conversion is low, typically 15%. The product gases pass through a condensing system, which raises steam, and condensed products are removed in a knockout pot. Steam reformers produce excess hydrogen relative to the stoichiometric requirement for methanol synthesis. This builds up in the methanol synthesis loop and has to be purged from the system and used as fuel.

Major operating costs in methanol synthesis are the synthesis gas compressor and the recycle compressor. Newer approaches aim to combine steam reforming with partial oxidation to produce a more suitable synthesis gas and reduce these compression costs.

The raw methanol from the knockout pot, also called *crude methanol*, contains large quantities of water and some higher alcohols. It is now passed to a fractionation train, where the required methanol grades are produced.

To a good approximation, the amount of water produced is proportional to the amount of carbon dioxide in the synthesis gas (reaction stoichiometry 8.2). For typical steam reformer synthesis gas, the water content is in the region of 15%–17% (by weight).

Higher temperatures favor higher alcohols, particularly butanols. Older high-pressure technology produced methanol containing several percent butanols. An old and now extinct technology (isosynthesis) produced high concentrations of isobutanol.

The overall efficiency of methanol production in stand-alone facilities is about 60%.[2] Quoted figures are often higher than this, but these are often on an *inside battery limits* (ISBL) basis. This basis only concerns the production of methanol and does not include gas used in the production of utilities. Further, higher figures do not account for typical operational losses.

Methanol grades. There are three principal grades of methanol. The key parameters for these grades are given in table 8–1.

Table 8–1. Methanol grades—key parameters

	Fuel	Grade A	Grade AA
Methanol (min. wt%)	>99%	>99.85%	>99.85%
Distillation range (C)		1(a)	1(a)
Water (max wt%)		0.15	0.1
Acetone + aldehydes (ppm)		<30	<30
Acetone (ppm)			<20
Ethanol (ppm)			<10
Residue (mg/L)		<10	<10

Note (a) must include 64.6C

- **Fuel methanol.** This can be produced in a single column. This product has water content reduced to < 1%. Since the presence of higher alcohols is not a problem with fuel grade, the product contains some ethanol. Higher alcohols are favored by high-temperature synthesis. Since higher alcohols have higher energy content, this could be advantageous for some end uses. There have been proposals to produce fuel methanol with several percent higher alcohol (particularly butanol) content. Fuel methanol is not traded but has been produced for various alternative fuel demonstration projects.
- **U.S. Federal Grade A methanol.** This can be produced in a two-column system. This grade is used when certain specifications in Grade AA are not critical, e.g., for the production of formaldehyde for the production of urea-formaldehyde resins.
- **U.S. Federal Grade AA methanol.** This is the grade generally traded. It has a high specification, particularly on the ethanol and acetone content. This grade is produced by a three- or four-column distillation system.

Downstream Products

Annual methanol production is about 30 MMt. Most use is for sophisticated chemicals and manufacturers. Figure 8–2 illustrates the split of uses.

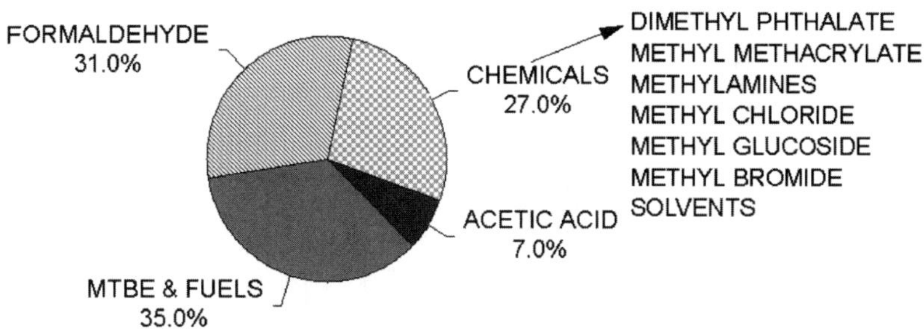

Fig. 8–2. Methanol uses

MTBE

The major use of methanol is for the production of methyl tertiary-butyl ether (MTBE), and to a lesser extent, as a fuel in its own right. MTBE is produced in refineries and petrochemical plants throughout the world from the reaction between methanol and isobutene.

$$CH_3OH + CH_2{=}C(CH_3)_2 = CH_3OC(CH_3)_3 \qquad (8.3)$$

There are several proprietary processes.[3] The principal features are shown in figure 8–3.

Excess methanol is added to a mixed olefin stream, typically containing between 12% and 30% isobutene (i-C_4= in the figure). The mixed olefin stream is sourced from either refinery fluid cat cracker (FCC) off gas or the C_4 stream of a naphtha steam cracker producing ethylene. The methanol-olefin mixture is passed to a reactor containing an acidic resin catalyst. This converts the isobutene in the feed to MTBE but leaves other olefins in the feed unaffected. The products are then passed to a distillation column that separates the MTBE from the excess methanol and C_4 raffinate. The methanol and C_4 raffinate are passed to a wash tower, where methanol is water washed out of the gas stream. A further distillation separates the methanol and the water, which are recycled.

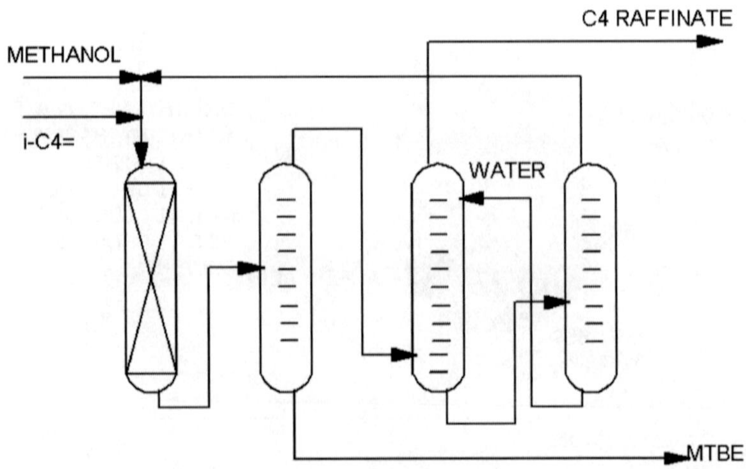

Fig. 8–3. MTBE synthesis

The feed temperature of the resin is typically near the boiling point of methanol (64°C). The reaction heat can be used to drive the distillation in an integrated system.

In the Middle East and Southeast Asia, MTBE is produced in large plants from the readily available butane-rich natural gas feedstock. The block flow for this process is illustrated in figure 8–4.

From large gas field development, such as for the production of LNG, a large volume of butane is usually produced. As an alternative to the LPG trade, this can be converted into MTBE. This is accomplished by first isomerizing the butane to increase the concentration of the branched isomer, which is then separated. The isobutane stream is passed to the dehydrogenation unit, which is the largest unit operation in this type of complex. There the isobutane is converted into isobutene. This, with methanol produced from some of the methane component of the natural gas, is then used to produce MTBE.

This ether is the preferred additive to automotive gasoline for increasing octane without resorting to lead or aromatics. It is also chosen to oxygenate gasoline for ameliorating the formation of photochemical smog from unburnt hydrocarbons and carbon monoxide emitted from vehicle exhaust. Unlike other additives, it is compatible with both gasoline and vehicle equipment.

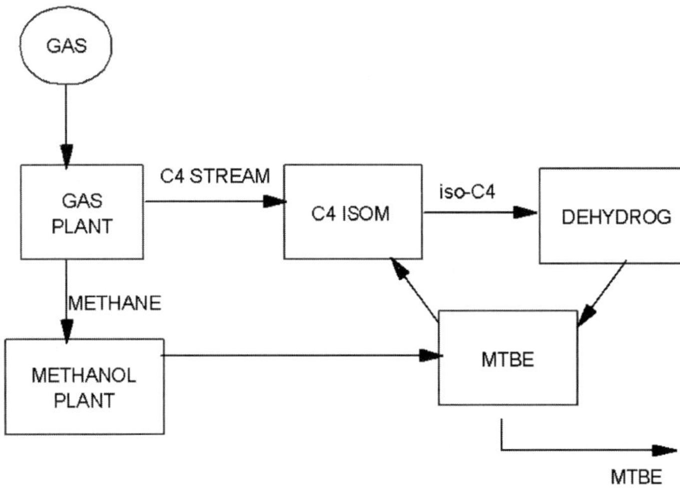

Fig. 8–4. MTBE from natural gas

Formaldehyde

Production of formaldehyde is the next largest use of methanol. This is used throughout industry mainly for the production of urea-phenolic and methylamine resins, which are used for the manufacture of chipboard and plywood. However, there are a wide variety of other uses of formaldehyde, such as in the manufacture of paints, cosmetics, explosives, fertilizers, dyes, textiles, and paper. Formaldehyde is made by the air oxidation of methanol using a catalyst.

$$CH_3OH + \tfrac{1}{2}O_2 = CH_2O + H_2O \qquad (8.4)$$

The reaction is very exothermic ($\Delta H = -236.5$ kJ/mol), and heat removal is central to controlling the reaction.

There are two production processes, one based on silver catalysts and the other on mixed metal (iron-molybdenum) oxide systems. The outline of the mixed metal process is shown in figure 8–5.

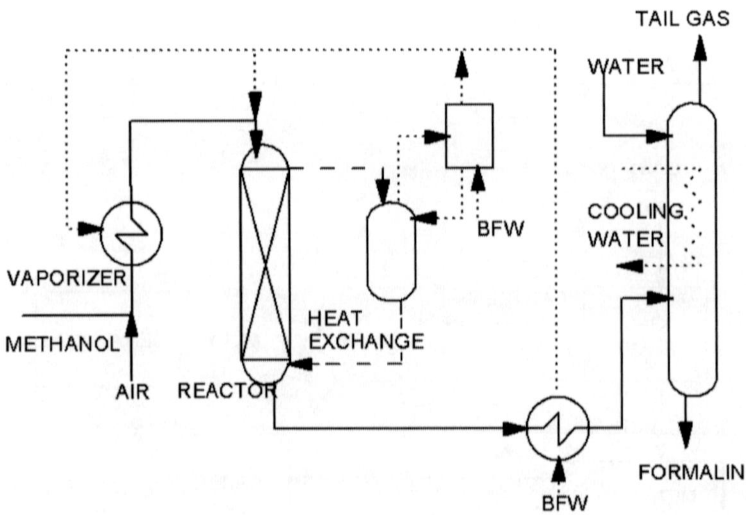

Fig. 8–5. Formaldehyde production

Methanol and air are passed to a vaporizer and then the reactor, where combustion occurs in the region of 300°C–400°C. There is extensive heat exchange producing steam by heating and vaporizing boiler feed water (BFW) in various parts of the process. After oxidation, the formaldehyde is washed out of the product gases and extracted as a water solution known as *formalin*. It is generally used in that form. The production of urea from natural gas is discussed in a later chapter.

Acetic acid

The third largest use for methanol is the production of acetic acid. The modern process was developed by Monsanto.[4] Methanol reacts with carbon monoxide in the presence of a rhodium catalyst and an iodine promoter.

$$CH_3OH + CO = CH_3COOH \qquad (8.5)$$

The layout of the principal operation is shown in figure 8–6. Carbon monoxide and methanol react in the homogeneous phase in a reactor. Product is removed and purified in a series of columns. Unreacted methanol and carbon monoxide are extracted and recycled.

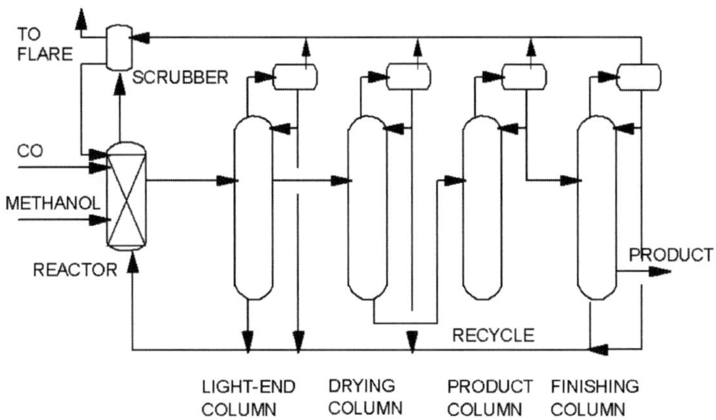

Fig. 8–6 Acetic acid manufacture—Monsanto process

Other chemicals

A major use of methanol is in the production of a wide variety of chemicals and solvents. The principal ones are:

Dimethylphthalate
Methylmethacrylate
Methylamines
Methylchloride
Methylglucoside
Methylbromide

Methanol Market and Prices

World methanol demand

The United States, the European Union, and Japan account for the major portion of methanol demand. For the most part, the underlying methanol growth will follow that of the world growth to GNP. However, demand has been influenced by the Clean Air Acts in the United States, which have forced the use of oxygenates, particularly MTBE. This resulted, during the early 1990s, in an increase in demand for methanol above the natural growth curve. Since 2003, controversy regarding the use of MTBE as a gasoline additive has had the potential to stall its growth. However, in Europe and in the Far East, there continues to be increasing demand for MTBE as a substitute for lead in gasoline in order to lift octane without resorting to the use of aromatics.

From a base level of 25 MMt/y, at an annual growth rate of 3%, this corresponds to 750,000 t/y, which is about the output of a world-scale plant each year. However, elimination of MTBE from the U.S. gasoline pool would be equivalent to 4 MMt of methanol demand removed from the market.

Operations. Methanol manufacturing operations can be considered in two parts. The first is the collection of generally small plants (< 1,000 t/d) in the countries using methanol. These are generally old plants with high operating costs. Many of these plants use feedstock other than gas (e.g., residual fuel or coal). They are protected by tariffs, such as in the European Union, or are on a care-and-maintenance basis until swings in the methanol price justify their operation. The second part consists of very large unit operations (>1,000 t/y) in countries with access to low-cost gas, such as Saudi Arabia, New Zealand, Chile, Canada, Trinidad, and Malaysia.

There are a few large producers, including Sabic (Saudi Arabia) and Methanex (Canada, with production facilities also in Chile and New Zealand). These principal players dominate the trade in methanol.

Chemical methanol prices

Figure 8–7 shows the historical spot price for methanol.[5] The figure illustrates the large price volatility for methanol ranging from lows of below \$100/t to highs approaching \$500/t. The figure clearly illustrates the trade-cycle nature of chemical methanol. Price peaks seem to occur regularly about every two to three years. The average price from the beginning of 1988 to year-end 2004 was \$166/t.

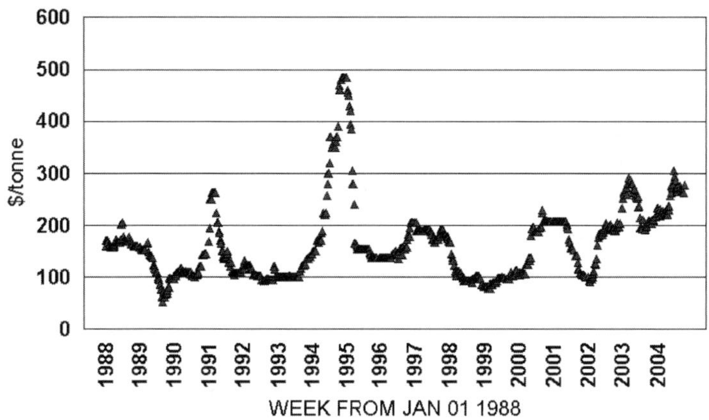

Fig. 8–7. Historical methanol prices

Analysis has been conducted of the prices in the major markets (Europe, Japan, and the United States).[6] This shows small price differentials (maximum of about \$50/t) that are rapidly eliminated by the international trade.

Chemical methanol shipping costs

Chemical methanol is transported in specialized chemical tanker ships in cargoes of 1,000 t or more. Chemical tankers are typically up to about 40,000 deadweight tonnes (DWT), and this class of ship is used for the transport of large parcels on transoceanic voyages. Prices for shipping chemical methanol are regularly reported by *European Chemical News* and its "Shipping Monitor" feature. Reported costs depend on the parcel size and distance travelled, typically ranging from \$10/t to \$25/t. Chapter 5 gives a further analysis of shipping costs.

Methanol Production Cost

The fixed variable relationship for the production of methanol is shown in figure 8–8. The estimates have been made from the statistics given in table 8–2.

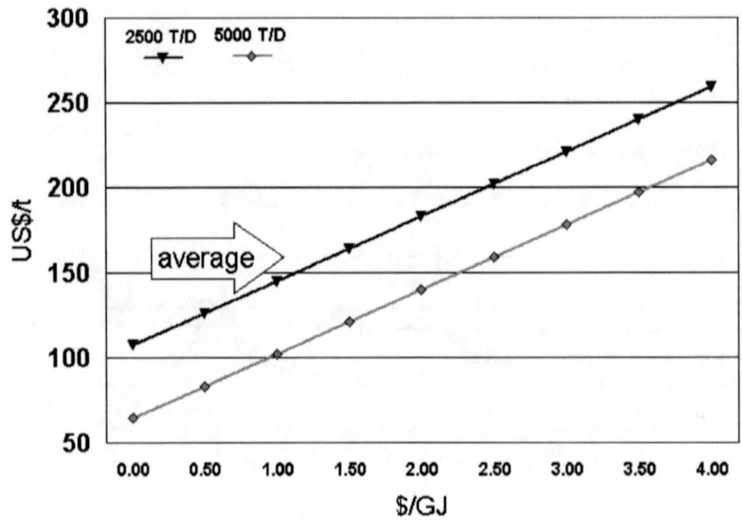

Fig. 8–8. Methanol production costs

Two cases are presented for the production of Grade AA methanol. The first case (2,500 t/d; 850 kt/y) represents a conventional technology based on a steam reformer. A steam reformer/secondary reformer would have broadly similar statistics. This case has an estimated capital costs in 2004 of $380 million. The second case (5,000 t/d; 1,700 kt/y) represents larger plant based on partial oxidation. These are claimed by several groups to have much lower capital costs, including all the utilities and the oxygen plant.[7] However, no plants of this design have yet been completed, although two are in the construction phase.

A comparison can be made between figure 8–7 and figure 8–8 concerning the average costs of methanol. This reveals that for a conventional plant, gas supply has to be below about $1.5/GJ in order to generate a production cost that is better than the average price.

Table 8–2. Statistics for methanol production

Production	kt/a	850	1700
Capex	$MM	378.6	452.1
Construction period	years	3	3
Plant life	years	15	15
Return on capital	%/a	16.34%	16.34%
Non gas operating costs	$MM/a	91.6	110.3
Gas usage	PJ/a	32.2	64.3

Most export plants are located in countries with considerably higher construction costs than the U.S. Gulf Coast. Thus the gas prices for a viable plant would be about $1/GJ or less.

Again using the long-term average price of methanol as a guide, the larger scale plant, in theory, could be viable at gas prices higher than $2/GJ.

Formaldehyde

After MTBE, formaldehyde is the major product that is produced from methanol. This is produced as formalin, a 37% aqueous solution. The cost of production can be estimated for a typical 40,000-t plant adjacent to a methanol facility at about $270/t. The pertinent statistics are given in table 8–3.

Table 8–3. Statistics for formaldehyde production

Production	kt/a	40000
Capex	$MM	12.29
Construction period	years	2
Plant life	years	15
Return on capital	%/a	15.49%
Non methanol costs	$MM/a	3.00
Methanol usage	t/a	45946
Total production costs	$MM/a	10.73
Unit production costs	$/t	268

Fuel Methanol and DME

Both of these materials are currently produced in small volumes, because they have at present limited uses. DME has a use as an aerosol propellant and is made in small amounts for this purpose. However, fuel methanol and DME have fuel properties that make them potential substitutes for gasoline and diesel fuel, respectively. Thus they may in the future be made on a very large scale.

Table 8–4 reviews the properties of fuel methanol and DME and compares them to propane and diesel fuel.

Table 8–4. Comparison of methanol fuels with propane and diesel

		Methanol	DME	Propane	Diesel
MW		32.04	46.07	44.09	170–200
Boiling Point	C	64.6	-25.1	-42	180–370
Liquid density (20C)	g/cc	0.79	0.67	0.49	0.84
Latent heat of vapor	kcal/kg	262.0	111.7	101.8	60.0
Vapor pressure @ 25C	bar	0	6.1	9.3	0
Auto-ignition temp.	°C	464	235	457	316
Explosion limits	Vol%	7.3–36.0	3.4–18.6	2.1–9.5	1.0–6.0
Octane number		106		96	
Cetane number		5	55 - 60	5	50
LHV	kJ/kg	20,060	28,430	46,360	41,800

Fuel methanol has similar properties to gasoline and can be stored and transported in a similar manner. Fuel methanol has a high octane value (> 100) and can be used in gasoline engines with minor modification as a neat fuel or as a gasoline/methanol blend.

It does have a low cetane value, but it can also be used as a diesel fuel by the addition of a cetane improver (methanol has a cetane of only about 5). Methanol has been extensively tested in both gasoline and diesel engines, and its beneficial properties over conventional fuels are well documented.

DME is more like LPG (such as propane) in its physical properties. It is stored and transported in a similar way in either pressurized vessels or by cryogenics. But by contrast, it is more suitable for use in a diesel engine due to its high cetane value.

These properties make fuel methanol and DME attractive for gas conversion to produce synthetic transport fuels.

Fuel methanol production

The production of fuel methanol from gas follows the same operation as for chemical methanol except for the final distillation steps. Fuel methanol can be made from crude methanol with a single distillation column.

The relatively simple production methods (small number of unit operations and well-known technology) raise the potential for using barge- or ship-mounted plants. This, in turn, raises the potential of low-cost construction at major fabrication yards and towing to remote areas to exploit low-cost gas.

There are now more than 30 years of experience in designing and developing the concept of offshore and floating production plants. Because of the simplicity of its production, most effort has concentrated on the production of methanol.

It is true that effort and interest has been focused on the construction and use of ocean-going barges and ships. However, there are large volumes of gas available in areas of shallow water depth. For such cases, barges could be used to float a production facility to the location. The barge could then be sunk, and the production would occur as a land-based operation.

However, despite all of this activity, there are no floating production operations known to the author. Changes to the technology over the years are summarized in table 8–5.

Table 8–5. Evolution of technology for offshore methanol synthesis

Issue	Solution	New Problem
Steam reformer size (ship stability)	partial oxidation	oxygen plant
Oxygen Plant		
fire hazard	fire walls	weight
stability	packed column:	new technology
product purity	PSA plants	low purity
Product Quality		
column stability	packed columns	increased weight
low purity product	low grade product purified onshore	higher overall costs
Unit operations	clean technology—no by-products	separate storage
Utilities and off-sites	all services to be provided on-board	space limitations

The earlier plants were based on steam reformer systems. The very large space requirement and size caused stability issues and needed extensive anchoring. The perceived solution was to use partial oxidation. This would either completely eliminate the steam reformer (partial oxidation) or greatly reduce the size of the reformer (e.g., GHR method). However, this generated the issue of the oxygen plant.

The production of oxygen in an offshore environment is extremely hazardous. Its production in close proximity to a hydrocarbon production facility would present extensive safety issues. This could be solved by the use of firewalls, but these would increase the overall weight of the topside facilities.

Another issue is the separation of oxygen by cryogenic distillation of liquid air on a moving ship. The common method onshore is by multiplated columns constructed to high tolerances. However, these would not necessarily perform well on a floating (moving) ship. This would cause changes to the oxygen purity that might upset the performance of the partial oxidation unit. The latter could be solved using packed columns, but this would involve new technology for cryogenic separation.

One solution would be to accept some oxygen purity loss and use PSA units to separate the air. These could be built in modules and stacked on deck. However, this reintroduces the issue of equipment size and sacrifices the economies of scale gained in using cryogenic plants.

Another issue for offshore production is the purity of the product. The normal approach would be to produce fuel methanol on a single packed column. However, production of Grade AA would be more problematical. BHP demonstrated the GHR technology in a small (120 t/d) plant in Australia. This incorporated two packed columns for producing methanol Grade A product.

Methanol has the advantage of being a single product. No other products would be produced. However, separate storage would be required if hydrocarbons (oil or condensate) were also to be produced by the facility. In addition, all utilities and ancillary services (off-sites) for the methanol plant would have to be provided, thus generating severe space constraints.

Fuel methanol production costs. Figure 8–9 compares the production cost for fuel methanol on a land-based facility and on a barge. The pertinent statistics are given in Table 8–6.

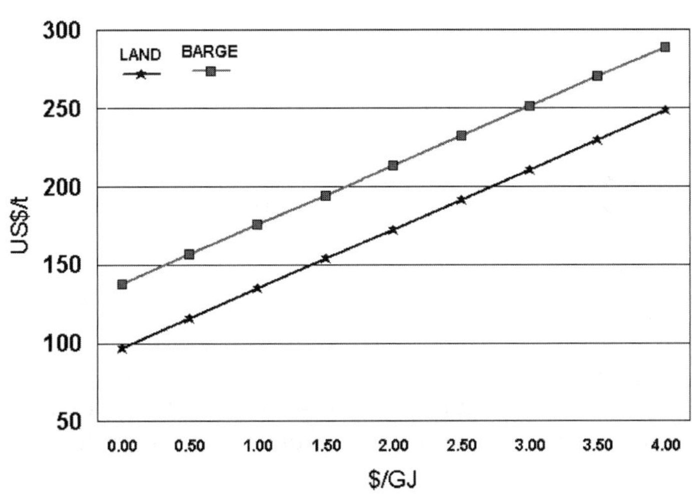

Fig. 8–9. Fuel methanol production costs

The capital cost of fuel methanol production is less than that for Grade AA, which is the most heavily traded grade, because of the much smaller fractionation train. In the offshore barge case, $146 million are allowed for the cost of the barge and for other ancillary services, such as the necessary gasfield infrastructure.

Table 8–6. Statistics for fuel methanol production

Production	kt/a	850	850
Basis		Land	Barge
Capex	$MM	340.7	486.1
Construction period	years	3	3
Plant life	years	15	15
Return on capital	%/a	16.34%	16.34%
Non gas operating costs	$MM/a	82.62	117.29
Gas usage	PJ/a	32.2	32.2

In practice, a land-based facility would purchase gas, whereas the barge facility would receive gas at a nominal price for the gas field over which it would sit. At $1/GJ, gas value for the barge facility is equivalent to a gas price of about $2.20/GJ on land, producing methanol at about $170/t.

Mass production of fuel methanol

For transport fuels, and potentially for olefins production, one must be concerned with the mass production of fuel methanol. For large gas reserves, many barge-mounted plants could be used. The construction of many barges and plants of identical design would be expected to generate considerable savings. Figure 8–10 and table 8–7 compare the production costs for a single barge and for 10 barges.

For the 10-barge case, each barge is estimated to cost 80% of the cost for a single barge. Further allowances are made for infrastructure (hook-up) and other ancillary services for the gasfield development. The figure illustrates that at a nominal $1/GJ, methanol production cost is about $130/t. Should the gas be considered of no value, the production cost would be about $100/t, or less than 12.5¢/L.

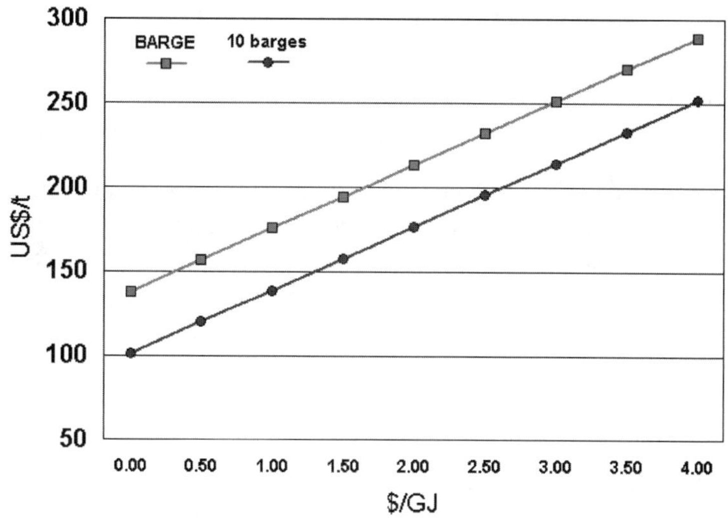

Fig. 8–10. Mass production of fuel methanol

Table 8–7. Statistics for mass fuel methanol production

Production	kt/a	850	8500
Basis		1 barge	10 barges
Capex	$MM	486.1	3543.4
Construction period	years	3	3
Plant life	years	15	15
Return on capital	%/a	16.34%	16.34%
Non gas operating costs	$MM/a	117.29	858.7
Gas usage	PJ/a	32.2	321.6

Dimethyl Ether (DME)

DME is receiving increased attention as an alternative fuel. The salient properties are presented above in table 8–4. The key points are:

- DME is a gas at ambient temperature and pressure and is therefore stored and transported at pressure, much the same as LPG.
- DME is proposed as a substitute for diesel, for which use it has beneficial environmental properties. Unlike methanol, it has a sufficient cetane number without the use of additives or spark assistance.
- DME has low octane and is not used as a gasoline substitute, in contrast to methanol, which has excellent properties for this use.
- The calorific value is higher than methanol but lower than conventional fuels.
- Like methanol, DME can be used directly in gas turbines.

DME production

DME is produced from methanol by the reaction:

$$2CH_3OH = CH_3.O.CH_3 + H_2O \quad (8.6)$$

The well-known process is very simple and is used to produce relatively small quantities of DME for various uses, such as a propellant in retail sprays. For the mass production of DME from gas sources remote from the major markets, there are two approaches. The first is a single process whereby DME is synthesized directly from synthesis gas without going through the intermediacy of methanol. The second is a two-stage process whereby methanol is produced first.

Single-stage route. Figure 8–11 illustrates the unit operations for DME synthesis and storage at a remote site followed by transport to a user. One should note that because DME is like LPG, storage and transport costs, although lower than LNG, are higher than the cost of storage and transport of conventional fuels. The higher costs relative to methanol would be offset by DME's higher calorific value. Hence similar-sized ships will carry more energy transporting DME than methanol.

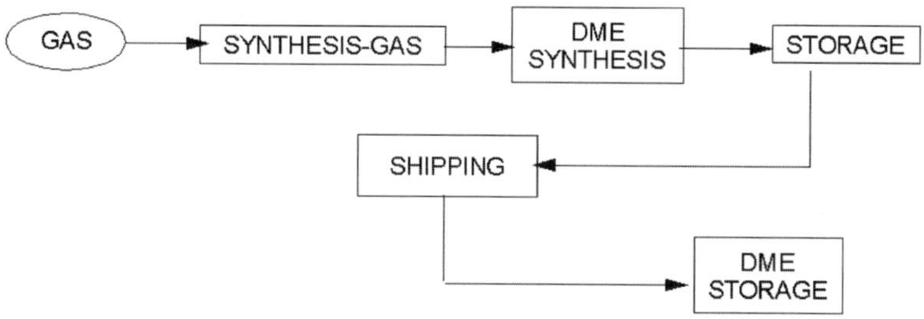

Fig. 8–11. DME single-stage route

There are questions concerning the technical viability of the single-stage route. The principal problems to date have been:

- Poor catalyst stability, with catalyst activity and selectivity falling away with time-on-stream. By comparison, methanol synthesis catalysts last many years.
- Requirement to reduce carbon dioxide in the synthesis gas for optimum performance. By contrast, methanol requires some carbon dioxide.

Several companies claim to have overcome these problems and offer a licensed single-stage technology. These are reviewed in table 8–8.[8] Note that some require synthesis gas of low stoichiometric ratio and produce a mixture of methanol and DME. These processes require a methanol dehydration unit if only DME is to be produced.

Table 8–8. Some licensors of DME technology

Licensor	NKK	Air Products	TOPSOE	TEC
H_2/CO	1	0.7	2	2
Reactor	Liquid Phase Slurry	Liquid Phase Slurry	Vapor Phase Fixed Bed	Vapor Phase Fixed Bed
Catalyst	Methanol Synthesis + Dehydration + Shift	Methanol Synthesis + Dehydration + Shift	Methanol Synthesis + Dehydration + Shift	Methanol Synthesis, then separate Dehydration
Temperature	250–320°C	250–280	210–290	
Pressure (bar)	30–50	53–102	70–80	
Products	>95% DME	30–80% DME	DME-MeOH mixture	

The method has been the subject of research over many decades because methanol can easily be obtained from DME by reversing the previous reaction:

$$CH_3.O.CH_3 + H_2O = 2CH_3OH \quad (8.7)$$

This offers the methanol producer potentially significant lower methanol production costs as a result of higher theoretical conversion. (One should recall that the pass conversion in methanol synthesis is only about 15%.)

Two-stage route. Figure 8–12 illustrates the pathway for producing DME in a two-stage route via methanol. The first stage (at a remote, gas-rich site) produces crude or fuel methanol. The methanol is then transported by tanker to another site near the product market, where the DME is produced. The DME is stored prior to shipment to the end users. This route uses proven technology at each step and offers the advantage of low-cost methanol transport in very large tankers.

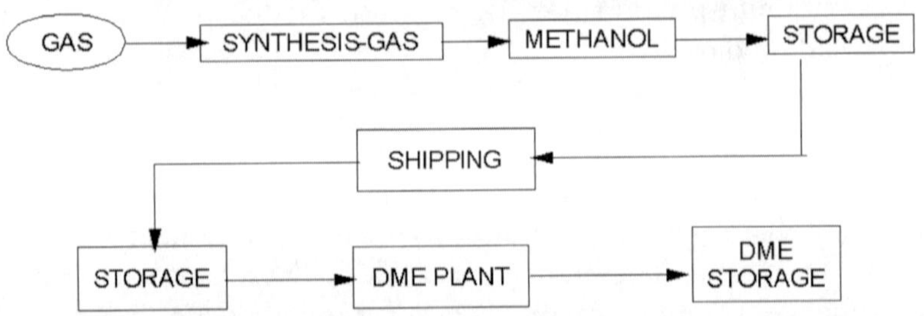

Fig. 8–12. DME via methanol

Production costs for DME

The conversion of crude or fuel methanol into DME is a simple process. The production costs of DME would be very similar to that for fuel methanol, even for the two-stage route. However, storage and handling costs would be higher, and an extra $50 million has been added for these costs. The fixed-variable relationship illustrated in figure 8–13 uses the statistics given in table 8–9.

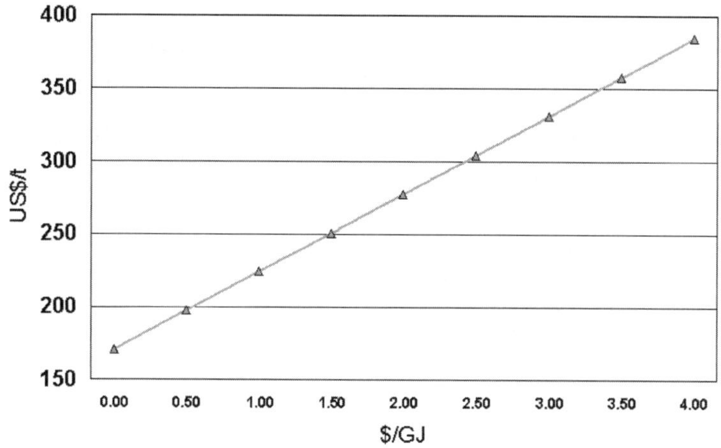

Fig. 8–13. DME production costs

Table 8–9. Statistics for DME production

Production	kt/a	610
Capex	$MM	390.71
Construction period	years	3
Plant life	years	15
Return on capital	%/a	16.34%
Non gas operating costs	$MM/a	104.18
Gas usage	PJ/a	32.61

Studies of Floating Methanol Production

There have been many proprietary studies for offshore methanol production. Some of these have been published.[9] Two are chosen that illustrate the types of projects considered.

Ocean barge plants. Table 8–10 illustrates the statistics for an ocean-going barge for production of methanol. Swedyards developed the concept during the 1970s using steam reforming to produce the synthesis gas. Detailed design of the methanol plant was conducted by Haldor-Topsoe. There were extensive modelling studies to prove the stability of the barge in crosswinds (note the 60-m high structure).

The concept was for the production of 3,000 t/d methanol (Grade AA) using offshore gas. No account was taken of associated condensate production that would improve the economics.

The barge is fitted with all the facilities required for the production of methanol, listed in table 8–11.

Table 8–10. Methanol barge statistics

Barge type	Rectangular—flat bottomed
Displacement	46,000t
Length	138m
Breadth	74m
Depth	20m
Draught loaded	7m
Draught ballasted	4.5m
Maximum height above deck	60m

Table 8–11. Statistics for a barge mounted methanol plant

Process Plant	Utilities	Other Facilities
desulfurization	steam raising	accommodation (75)
steam reforming	sea water de-salting	buoy
heat recovery	cooling water	bilge
methanol synthesis	inert gas	product storage
methanol distillation	instrument air	safety gear
	electricity	

FPSO-mounted methanol plant. The case described was developed in the early 1980s. It utilized associated gas produced during the production of crude oil on a floating production storage and off-take (FPSO) tanker. This overcomes the flaring of gas, which may be prohibited or taxed. This concept described a small 900-t/d methanol plant (compared to a world-scale plant, 2,500 t/d), with oil production of 60,000 bbl/d oil. The statistics are illustrated in table 8–12.

The FPSO tanker is equipped with the services to produce both methanol and crude oil (table 8–13).

Table 8–12. Statistics for methanol on FPSO

Ship type	FPSO central turret
Displacement	87,000t
Length	269m
Breadth	46m
Depth	21m
Draught Loaded	15m

Table 8–13. FPSO facilities

Process Plant	Utilities	Other Facilities
steam reforming	steam raising	accommodation (80)
heat recovery	sea water desalting	buoy
methanol synthesis	cooling water	bilge
methanol distillation	inert gas	methanol and oil storage
	instrument air	safety gear
	electricity	
	crude oil recovery	

References

1. Twigg, M. V., editor. *Catalyst Handbook*, Wolfe Publishing Ltd., 1989; S. Lee. *Methanol Synthesis Technology*, CRC Press, 1990; and Cheng, W.-H., and H. H. Kung, editors. *Methanol Production and Use*, Marcel Dekker, 1994.

2. Cheng and Kung, editors. *Methanol Production and Use*, 1994.

3. For example, Axens NA, CD Tech, Uhde Edelenau. "Refining Processes 2002," *Hydrocarbon Processing*, November 2002, pp. 106–108.

4. Kung, H. H. and K. J. Smith. "Methanol to Chemicals," Methanol Production and Use, referencing H. D. Grove, Hydrocarbon Processing, November 1972, p. 76.

5. Reported weekly in *European Chemical News*.

6. Regional prices are reported by Technon-Pirelli.

7. Davy Process Technology/Synetix UK and Lurgi Oel·Gas·Chemie GmbH. "Petrochemical Processes 2003," *Hydrocarbon Processing*, pp. 104–105.

8. Kato, K. "Future Outlook for GTL Fuels—A Japanese Perspective," *Hydrocarbon Asia*, July/August 2002, p. 24.

9. Ostby, M., and N. Nystad. "Technical/Economic Feasibility Study for a Floating 1000 M. Tons/Day Methanol Plant in the North Sea," Society of Petroleum Engineers, Houston meeting, 1978, SPE Technical Paper 7401.

9

METHANOL GTL DERIVATIVES—MTG, MTO, AND MTP

The last chapter dealt with the conversion of gas into methanol and into fuel methanol and DME. The latter two have been proposed as the products for large-scale development of gas reserves. This chapter will deal with methanol derivatives, which are hydrocarbons in nature. They are produced by the catalytic conversion of methanol or DME (using molecular sieves) into a hydrocarbon product and water.

$$nCH_3OH = -[CH_2]_n- + nH_2O \tag{9.1}$$

where $-[CH_2]_n-$ represents a general hydrocarbon product.

One of the products, gasoline, has been commercialized. Others that produce olefins or polymers are near commercialization.

For methanol to gasoline (MTG), a major plant built in New Zealand in the 1980s is analyzed. To further utilize gas, the production of olefins from methanol (methanol to olefins, MTO) has been extensively researched. The prospects for this route to olefins is discussed.

Overview of molecular sieve conversion processes

The conversion of methanol, produced from natural gas, to gasoline was practiced on a commercial scale in New Zealand during the 1980s. The process is commonly known as methanol to gasoline, MTG. MTO and methanol to propylene (MTP) have been demonstrated on a semicommercial scale in Europe. These technologies rely upon the unique shape-selective properties of molecular sieves. One such sieve is zeolite ZSM-5, which was discovered by Mobil Oil Corp. in the 1960s. Zeolites form a family of molecular sieves based on silica and alumina. Similar molecular sieves are based on alumina and phosphates (ALPO), and others on silica, alumina, and phosphates (SAPO). Molecular sieves differ from each other in the channel sizes they contain and in the nature (acidity) of catalytic sites. They thus discriminate between what they let into the sieve and what they let out. Acid centers within the channels catalyze the reactions on classical acid-catalyst terms. The basis of conversion of methanol by molecular sieves is illustrated in figure 9–1.

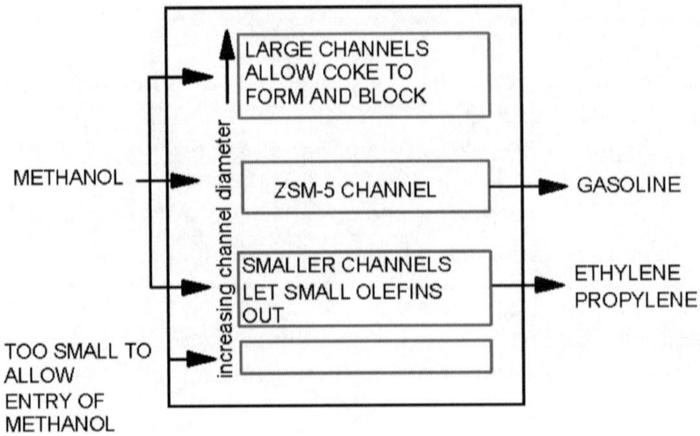

Fig. 9–1. Basis of molecular sieve catalysts

Methanol can enter sieves with channels over a certain specific size, for instance, those channels with a diameter larger than the molecular dimensions of methanol. Some sieves have no suitable channels for methanol. Some of these small port sieves (e.g., zeolite A) are capable of absorbing molecules smaller than methanol, such as water. They are used for drying or separating gases.

When the channel size is very large, and if the sieve has acidic sites within the channels, then unrestricted conversion of methanol can occur. This proceeds to form coke, resulting in the early fouling of the catalyst. This is exploited in the conversion of large heavy molecules in refinery fluid cat cracking. Zeolite Y (a large port zeolite) allows the entry of large molecules, breaks them into smaller molecules on acidic sites, and removes excess carbon as catalyst coke.

Critically, the size of the channel in ZSM-5 allows the entry of methanol and egress of molecules not much larger than xylene. It will produce products that boil in the gasoline range, free of poly aromatics, which would otherwise form coke. This is the basis of the MTG process.

It is useful to consider that the conversion of methanol, using this zeolite as a catalyst, proceeds first to form olefins. These are subsequently converted to aromatics and isoparaffins, the principal components of MTG gasoline. This picture is a crude approximation since it is known, for instance, that aromatics can be involved in olefin formation.[1]

Light olefins such as ethylene and propylene are the basic building blocks for polymers and have high economic value. There has been much effort directed towards the conversion of methanol to olefins per se. One method is to dope the zeolite catalyst with various oxides (such as phosphorus, zinc, or magnesia).[2] Another is to add water to the methanol feed by or operating with ZSM-5 at low pass conversion.[3] Other methods include the use of smaller port zeolites. While allowing the entry of methanol, these smaller port zeolites only permit the egress of small olefins—ethylene, propylene, and n-butene.

Low pass conversion using ZSM-5 was used in a large-scale demonstration undertaken by AECI in South Africa. They developed a process to pilot scale operating at low per-pass methanol conversions in order to maximize the olefin yield. Yields of ethylene of more than 35% of the hydrocarbon product have been obtained.

In West Germany, the UK Wesserling demonstration plant has shown that large quantities of olefins, particularly propylene and higher, can be extracted from a MTG designed plant. This plant is based on a large fluid-bed reactor.

The fluid-bed technology can offer advantages over the fixed-bed reactors. These advantages include a smaller reactor, use of methanol as opposed to DME as feed, and much improved steam raising ability. It also seems likely that the fluid bed can offer more flexibility to produce olefins. The olefins could be an intermediary feedstock for a polymerization plant producing distillate—the so-called MOGD process.

Mobil Methanol to Gasoline (MTG)

The commercial MTG operation built in New Zealand during the 1980s used fixed-bed reactors to convert the methanol into gasoline. Gasoline has a lower intrinsic hydrogen to carbon ratio than the hydrocarbon portion of methanol, which can be regarded as $CH_2.H_2O$. The hydrogen balance is achieved by the formation of large quantities of LPG (propane and isobutane). This is then used to fuel the steam reformers that will produce the methanol. As an alternative, the LPG could be used as a feed to an alkylate plant to produce more gasoline.

The political history has yet to be written behind the New Zealand government's decision to build a natural gas to synthetic fuels plant based upon the Mobil MTG technology. The decision was momentous, since it committed considerable finance to the building of a full-scale plant based on a technology that at the time was proven only on a small pilot plant. The following comments are based on views and opinions expressed to the author on the origin of the venture.

The key to the technology, the zeolite ZSM-5, was developed internally within Mobil laboratories. Mobil laboratories also pioneered its use for certain petrochemical processes, in particular xylene isomerization. During the 1970s, the use of the Mobil zeolite completely revolutionized most xylene isomerization plants throughout the world. The revolution in this technology created the opportunity for large royalty and licensing fees to be paid for the zeolite. It seems commonly accepted that Mobil, the technology discoverers and owners, developed the MTG process with the aid of U.S. Department of Energy (DOE) grants. These grants were given during the years following the oil price shock of the early 1970s. Much of this early work, excepting catalyst synthesis and fabrication, is available in DOE reports.

It is also widely believed that, until the approach of the New Zealand government, Mobil was about to shelve MTG. It was instead going to concentrate all efforts on petrochemical processes, where higher royalties could be charged.

Before deciding to go with MTG, the government of New Zealand compared the proposed MTG with FT processes. The decision was made to develop the Mobil process to full scale, including payment for a demonstration of the technology on a larger plant in the United States.

Plant flow sheet

An outline of the plant flow sheet is shown in the figure 9–2. The MTG plant in New Zealand comprises two large steam reformers and methanol synthesis units, each with a capacity of 2,500 t/d of methanol. Because the intention was to convert the methanol into gasoline, the crude methanol produced was not purified to chemical grade methanol, which would have been suitable for export. The new owners (Methanex) have since installed this capability. The crude methanol is then converted to DME, which helps energy management in the downstream reactors, and then into gasoline and LPG in fixed-bed converters.

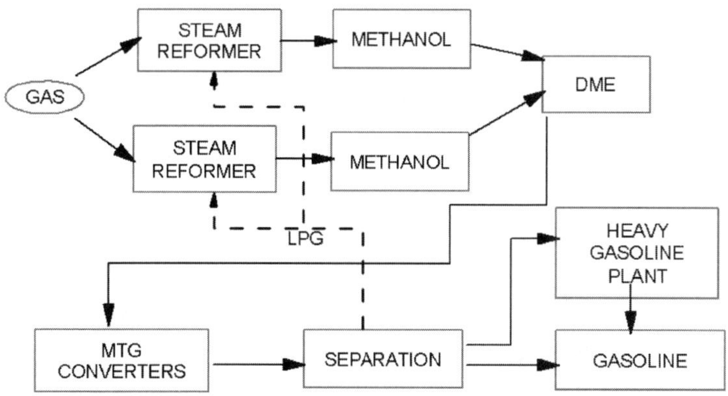

Fig. 9–2. MTG unit operations

As originally conceived, extra gasoline was to be produced from the LPG fraction by incorporation of an alkylation plant. However, this was not added, and the LPG was used to fuel the steam reformers. The operation of the MTG converters was then optimized to maximize the quantity of the gasoline produced. This was achieved by variation in reaction variables, such as temperature and space velocity. However, it also tended to increase the amount of durene (1,2,4,5-tetramethyl benzene) to about 4% or 5% in the product. Durene is troublesome in some vehicles, because its high melting point (80°C) results in carburetor blockages. The crude gasoline from the MTG converters was fractionated. The durene was removed in a heavy gasoline treatment plant by hydrogenation that resulted in the conversion of the durene to smaller aromatic molecules. This hydrotreated product was added to the light gasoline to produce a gasoline for immediate use.

Methanol to gasoline and distillate (MOGD)

The MTG process produces no distillate. A variation of MTG, the Mobil MOGD process, again uses a zeolite catalyst. It operates at lower temperatures and higher pressures, producing both a gasoline and a diesel fraction. Although highly branched isomers may be formed within the zeolite, only the less branched (hence higher cetane number) may escape through the zeolite pore window. Claimed cetane numbers are about 50, which is adequate for a motor distillate pool. However, rates of oligomerization using zeolite catalysts are lower than conventional systems, and fouling of the pores is a serious drawback. It is claimed that the MOGD process has been demonstrated using refinery produced olefins, but little information is available.[4]

Methanol to Olefins (MTO)

MTO processes are variants of the MTG process. The various options for producing olefins from natural gas via methanol are illustrated in table 9–1. These are compared with the Synthol process. This process is a variant of the FT process used in South Africa at Mossel Bay to produce an olefinic product from gas. In the designs of the existing operations, most of the olefin streams were subsequently converted into gasoline and diesel rather than being a source of olefins per se. However, the size of the operation and the quantity of olefin product (25%) in the light fractions make the Synthol-type operation a potential source for both olefins and fuels as separate products.

Table 9–1. Alternative approaches to olefin production

	MTG	MTC	UC MTO	UC MTO	MTP	SYNTHOL
Ethylene	3.2	25.2	45.6	33.6	0	4
Propylene	4.7	16.5	29.6	44.6	67.9	11.4
Butenes	8.3	5	9.5	12.8	0	9.3
Total olefins	16.2	46.7	84.7	91	67.9	24.7
Fuel gas	21.2	15.6	5.6	2	6.1	17.8
C5–160C	58	33	5.5	5.5	26	32.6
160–350C	5	1				13
>350C						5.4
Other losses (water phase or coke)		3.7	4.2	1.5	0	6.5

Early Mobil MTO processes

Early attempts to convert methanol into olefins were based on the zeolite ZSM-5. The Mobil MTO was based on the fluidized-bed version of the MTG technology. Conversion took place at about 500°C, allegedly producing almost complete methanol conversion. However, careful reading of the patent literature indicates that complete methanol conversion may not have been achieved by this means. Because of incomplete conversion, it would be necessary to strip methanol and DME from water and hydrocarbon products in order to recycle unconverted methanol.

In this variant, total olefin yield accounted for less than 20% of the products, of which ethylene was a minor but not insignificant product. The major product was gasoline. Ethylene is difficult to process and had to be treated specially. Claims that the ethylene could be recycled to extinction conflict with the known behavior of ethylene in zeolite catalyst systems and have to be viewed with suspicion.

The MTC process

The MTC process was primarily designed to produce ethylene by operating an MTG-type catalyst and process at low pass conversion in a fixed-bed reactor. The route has been developed by AECI, which demonstrated the process to pilot plant scale.

The principal reaction is brought about at low conversion in a series of reactors (10% conversion per reactor with about 40% conversion overall). The products, both aqueous and hydrocarbon phases, are heavily laden with methanol and DME. As a consequence, extensive extraction and recycling is required.

The principal product is ethylene, which is a valued petrochemical. The higher products are rich in olefins (66% olefins in C_3 and C_4, which are 41% of total). Like MTO, this process also produces a good quality gasoline and a heavy gasoline, which may require hydrotreatment.

UOP MTO process

The Union Carbide process, developed jointly with Norsk Hydro/Statoil, has been developed to semicommercial scale in Norway. The process uses proprietary catalysts based on a SAPO molecular sieve.

Two variants of the process are available, one maximizing ethylene and the other propylene. (Data for both are given in table 9–1). The performance appears to be similar to that of the conversion of methanol to olefins using small pore zeolites. Such systems suffer from high methane yield, which has to be recycled back to the reformer, and high coke yields. The formation of olefins is promoted by using crude methanol, which contains about 17% water.

The coke formation leads to catalyst fouling. This is solved in the UOP process by continuously removing a portion of the catalyst and passing this to a separate regenerator. After regeneration by combustion of the coke in air, the catalyst is sent back to the main reactor. In concept, this is similar to fluid cat cracking of refinery stocks. The process layout is illustrated in figure 9–3.

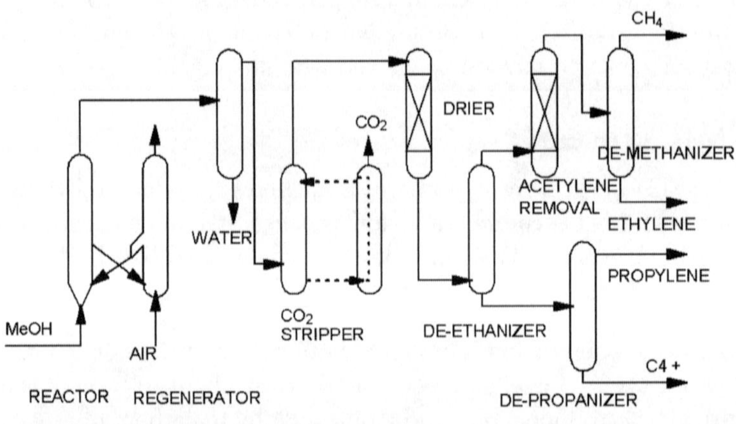

Fig. 9–3. UOP methanol to olefins

After separation of the mixed olefins, the product workup is similar to that in a steam cracker using LPG feedstock. Carbon dioxide is removed, and the hydrocarbon gases are dried before passing to a de-ethanizer column. The C_2 fraction is passed to an acetylene removal unit before methane is removed from the C_2 stream, which is 98% or more ethylene. The remainder is ethane. The C_{3+} stream is split between the C_3 fraction (98% propylene) and C_{4+}. The workup of the C_4 stream to produce linear butenes (not shown in the figure) is likely to be less problematic than the corresponding C_4 stream from steam crackers. The latter is highly complex and cannot be separated by fractionation alone. The process produces little product above C_5.

The Lurgi MTP process

The process has been demonstrated on a pilot scale by Lurgi and Statoil. Sufficient propylene has been produced to make polypropylene resin product by Borealis.[5] This process appears to use an oxide-doped ZSM-5 zeolite catalyst in fixed-bed reactors. The oxide doping promotes the methanol conversion to olefins. All olefins other than propylene are recycled to extinction, purged as fuel gas, or produced as naphtha. The flow sheet is illustrated in figure 9–4.

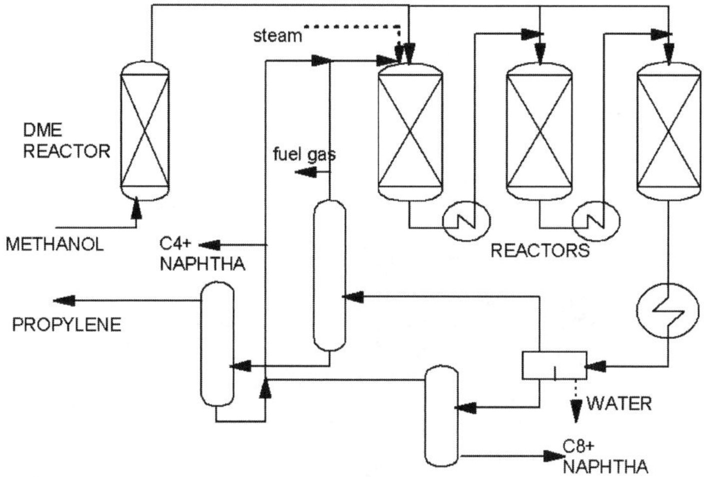

Fig. 9–4. Lurgi methanol to olefins process

Because fixed-bed reactors are used, the heat of reaction must be removed. This is achieved by initially converting some of the methanol to DME in a first reactor (similar to MTG). The next step involves splitting the feed to a series of reactors. Overall, the method resembles the operation of a methanol quench converter, where fresh feed is introduced at different points within a single reactor. Operation is at about 500°C, at which temperature propylene is favored over ethylene. Overall promotion of olefin yield is obtained by adding steam. Downstream of the reactors are separation columns that separate the C_3 product (about 80% propylene) from naphtha and fuel gases.

Production Costs

MTG—New Zealand plant costs

Because of the government involvement in the New Zealand Synfuels project, there is a relatively large amount of cost data for the project in the public domain. The New Zealand situation is often quoted in papers when considering the possibility of gas-conversion technology. It is important, therefore, to note the published cost breakdown. The total cost (in 1985 U.S. dollars) was $1,475 million. The cost breakdown is summarized in figure 9–5.

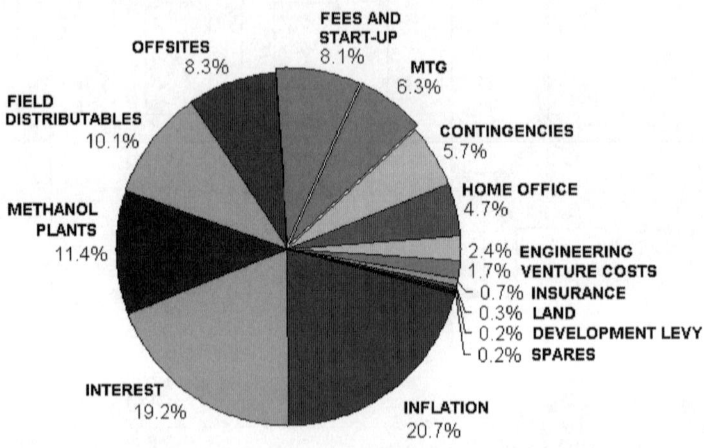

Fig. 9–5. New Zealand MTG cost breakdown

From this figure it can be deduced that the cost of the plant in terms of plant hardware only represents 53.6% of the total. These costs are inflated because of extensive field engineering, since the original site was thought unsuitable. The present site suffered ground subsidence, and extra piling was provided. The geological instability of the area has demanded the plant be able to withstand severe earthquake shock, as the plant is in sight of an active volcano. The item for capitalized engineering is thought to be payment to Mobil for research and demonstration specifically for this plant.

However, as illustrated, nearly 50% of the project's final cost is the result of fiscal charges (e.g., inflation and interest). Further, the published charges do not add up to the total quoted for the project of US$1,475 million. The costs detailed in the figure are from two sources.[6]

Estimated MTG production costs

The New Zealand MTG plant, located at Motunui on the north island, was designed to produce 13,000 bbl/d of gasoline from natural gas. As discussed above, the plant can be regarded as consisting of methanol synthesis and gasoline synthesis. In this case, the methanol synthesis uses well-tried technology (ICI, with Davy McKee as engineers). The latter uses Mobil technology with fixed-bed reactors rather than the now-preferred fluid-bed reactors demonstrated by UK Wesserling. These were not available at the time of plant construction.

Taking these comments as a basis, three scenarios for MTG production are evaluated for a 2004 GTL plant using this technology. The first is a high-cost case that takes the New Zealand data and scales the cost to 2004 values. The very high capital cost has forced some other changes in methodology. Working capital is taken as 30 days of stock at a value of $35/bbl. The labor, maintenance, and other charges were reduced in line with the fixed plant (50% of the total).

The second case considers the influence of capital charges on the production costs. This is done by taking a longer term view of the investment, hence requiring a lower return on capital. The case considers a 4-year construction phase, with a plant operating for 30 years at a DCF rate of 7.5%.

The third scenario is a low-cost case, with capital at $850 million. This is based on two crude methanol units at $300 million each and $250 million to cover methanol conversion and gasoline treatment. This scenario also envisages a 3-year construction period as opposed to the 4-year construction period of the other two cases. Interest rates are for the base case: 10% DCF, 15-year plant life. The fixed-variable relationship for the three cases is illustrated in figure 9–6, and the statistics are summarized in table 9–2.

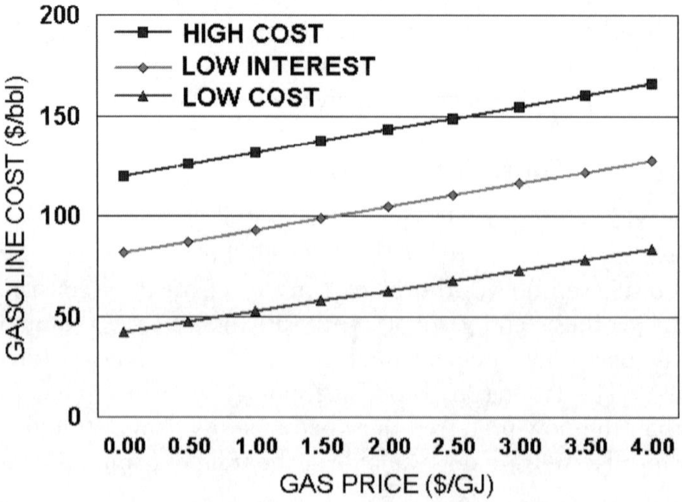

Fig. 9–6. MTG process economics

Table 9–2. Statistics for MTG production

Production	kt/a	516.5	516.5	570
Capital cost	$MM	2548.164	2548.164	850
Construction period	years	4	4	3
Plant life	years	15	30	15
Return on capital	%/a	0.169	0.103	0.163
Non gas operating costs	$MM/a	530.794	361.437	205.985
Gas usage	PJ/a	50.66%	50.66%	50.66%

The figure shows that at any gas cost, the MTG process cannot be viable using the New Zealand experience as a basis. Using a lower capital cost basis, gas at $1/GJ will produce gasoline at about $50/bbl. This is equivalent to crude oil at about $40/bbl. Note that this cost is well below the resale price of gasoline in many countries where transport fuels are heavily taxed.

MTO Costs

No full-scale MTO plant currently exists. Thus it is necessary to build up the economic statistics from what is known about the outcome of studies and from the analysis of the MTG process given above. The basis of the analysis is to consider the olefins that could be made from the output of two world-scale methanol plants with a capacity of 2,500 tpd crude methanol, 1,700 kt/y methanol. For olefins, Grade AA methanol is not necessary, since water generally improves olefin yield. Crude methanol (containing about 17% water if made by the standard steam-reforming route) would suffice.

Heat evolution is an issue for the conversion of methanol into olefins, and this is mitigated by conversion of a portion of the methanol into DME. This is analogous to the front end of the fixed bed MTG process. From this point, two options (UOP/Hydro MTO process and the Lurgi MTP process) will be further discussed.

UOP/Hydro MTO process

This MTO process seeks to convert crude methanol into ethylene, propylene, and butenes. The other product is a C_{5+} light naphtha. Fuel gas is used in the methanol synthesis operations, and the gas demand is reduced appropriately. Workup of the raw hydrocarbon products from the converter is by fractionation, similar to that for conventional steam cracker product separation.

The current flow sheet contains an acetylene hydrogenation unit on the ethylene stream. This was not present on the earlier reported versions of this technology.[7]

Most of the capital cost of $900 million comes from the cost of the two crude methanol plants at $300 million each. The remaining $300 million cost is for production of olefins to polymer grade specifications and provision for intermediate storage and off-sites for product handling.

The MTO process can deliver varying ratios of ethylene to propylene. For simplicity, the process is modeled in high ethylene/propylene ratio version. For ease of analysis, propylene is assumed to have the same economic value ($/tonne) as ethylene. The butene stream is assumed to have the economic value of naphtha.

MTP process

The proponents of the process claim MTP has been demonstrated on a plant juxtaposed to a large methanol plant in Norway.[8] No detail is given on the scale of the demonstration or the extent of the recycling operations. Propylene was shipped elsewhere for conversion into products. No detail is available on the amount of propylene processed. Since little detail is known of the MTP process, the analysis is based on the cursory information disclosed in patents.

In the flow sheet (fig. 9–4), only propylene and naphtha are produced as salable products. Fuel gas is assumed to be consumed in methanol synthesis.

The capital cost at $850 million is estimated at slightly less than that for MTO because there is no requirement to extract and purify either ethylene or butenes. However, this cost savings will be offset to some extent by the costs of handling the various recycling streams. The capital is made up of methanol plants (two plants at $300 million each) and $250 million for product workup and recycling.

Process economics

The fixed-variable relationship for the two cases (MTO and MTP) is illustrated in figure 9–7, and the statistics are summarized in table 9–3.

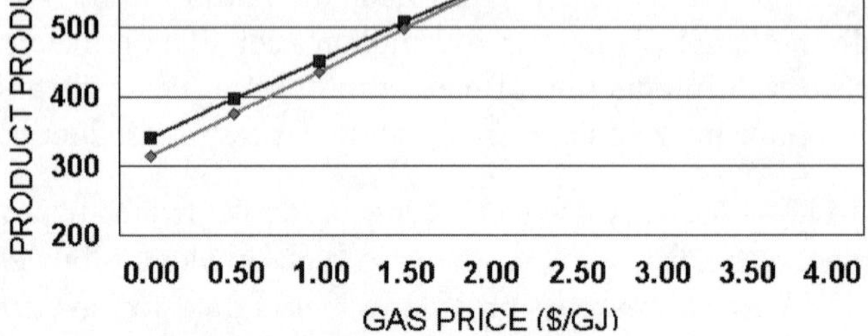

Fig. 9–7. MTO/MTP production economics

It can be seen that both MTO and MTP have broadly similar outcomes in terms of production costs. There is considerable volatility in the price of olefins, which are the subject of trade cycles. This is illustrated in figure 9–8.

The graph shows the spot prices on the European market for ethylene, propylene, and naphtha. The commodity olefins, ethylene and propylene, are traded at typically \$500/t, with prices ranging from less than \$300/t to more than \$800/t. The price differential between ethylene and propylene varies geographically and in the past decade has changed to favor propylene. This aspect is ignored in this analysis, and both olefins are assumed to have an equivalent value of \$500/t.

Table 9–3. Statistics for MTO-MTP production

		MTO	MTP
Production	kt/a	559.30	505.01
Capital Cost	\$MM	900.00	850.00
Construction period	years	3	3
Plant Life	years	15	15
Return on Capital	%/a	16.34%	16.34%
Non gas operating costs	\$MM/a	218.87	206.71
Gas usage	PJ/a	62.10	61.92
By-product credits (a)	\$MM/a	27.89	48.34

Note (a) butene and naphtha @ \$250/t

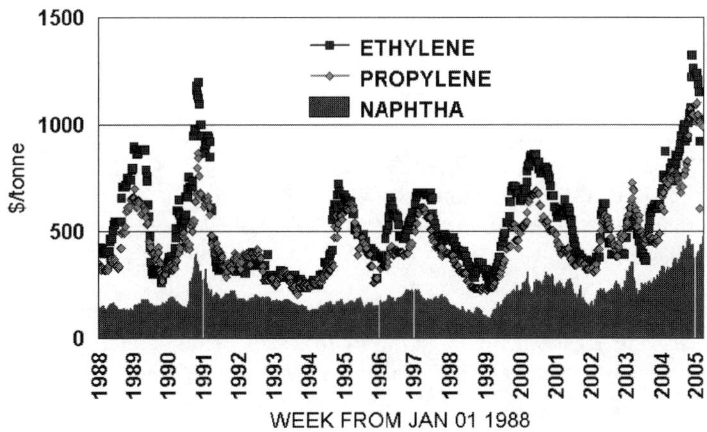

Fig. 9–8. Olefin and naphtha prices. Data from Market Report of *European Chemical News*.

The price of naphtha is also included in the figure. This is a by-product of the MTO and MTP processes. It is the principal feedstock for the conventional method for producing olefins, namely naphtha steam cracking. The value of naphtha is directly related to the prevailing price of crude oil. A value of $250/t has been used in this analysis.

Using this value, olefin production should be competitive with conventional product at gas prices less than $1.5/GJ. However, it should be noted that the ethylene and propylene trade is dominated by pipeline transfers and local shipping in the major markets of the United States, Europe, and North Asia. There is little trade outside this. If olefins are to be the product, then the relatively high cost of shipping (typically $40/t for propylene and about double this for ethylene) needs to be taken into account.

One obvious approach would be to integrate the olefins production with that of resin manufacture (i.e., polyethylene and polypropylene). This option increases the complexity and capital cost of the operation and is beyond the scope of this work.

Neither one of these technologies has been commercialized, and the projected output and costs may be rather sanguine. Along with the other issues discussed, it would seem that these technologies would require gas to be available at less than $1/GJ.

References

1. Mole, T., J. A. Whiteside, and D. Seddon. "The Effects of Aromatics on Methanol Conversion over Zeolite Catalysts," *Journal of Catalysis,* (82) 261, 1983; and T. Mole, G. Bett, and D. Seddon. "Conversion of Methanol to Hydrocarbons over ZSM-5 Zeolite," *Journal of Catalysis,* (84) 435, 1983.

2. McIntosh, R. J., and D. Seddon. "The Properties of Magnesium and Zinc Oxide Treated ZSM-5 Catalysts for Conversion of Methanol to Olefin-Rich Products," *Applied Catalysis,* (6) 327, 1983.

3. Seddon, D., T. Mole, and J. A. Whiteside. "Hydrocarbon Synthesis from Methanol—Using Zeolite Catalyst and Hydrocarbon Promoter," WO 8201866; and D. Seddon and T. Mole. "Conversion of Methanol Feedstock to Hydrocarbon Compounds—Using H-ZSM-Zeolite, Forming High Proportion of Ethylene," Australia Patent No. 8285988.

4. Tabak, S. A., and F. J. Krambeck. "Shaping Process Makes Fuels," *Hydrocarbon Processing*, September 1985, p. 72; and W. E. Garwood. "Conversion of C_2–C_{10} to Higher Olefins over Synthetic Zeolite ZSM-5," ACS Symposium Series 218, p. 383, and references therein.

5. Borealis A/S. "First Polypropylene Product Made from Natural Gas," Press release, September 16, 2003.

6. Bibby, et al., editors. "Methane Conversion," *Studies in Surface Science and Catalysis* vol. 36, Elsevier, 1988; and *New Zealand Synfuel*, Auckland: Cobb/Horwood Publications, 1985.

7. Compare UOP flow-sheets in "Petrochemical Handbooks," *Hydrocarbon Processing,* 1997 and 2001.

8. Hansen, R., and O. Olsvik. "Norwegian Methanol Plant Outlines Operations, Expansion Plans," *Oil & Gas Journal*, February 7, 2000, p. 46.

10

Gas to Liquids (GTL)— The Fischer-Tropsch Process

This chapter will consider the conversion of gas to hydrocarbon products typically produced in the oil industry. These include synthetic crude oil, distillates, and hydrocarbon chemicals, and the technology is commonly referred to as GTL technology. However, note that sometimes GTL is used to mean gas to methanol (a liquid); this was dealt with in an earlier chapter.

The interest in developing gas fields (so-called stranded gas) has led to increased interest in this process route. At the end of 2002, 14 projects were in the planning or engineering stages around the world. However, in 2005 there were only 8, with 1 (Oryx GTL Ltd. at Ras Laffan) under construction.[1]

In the conversion of gas into hydrocarbon products, the process used is the Fischer-Tropsch (FT) reaction. This term is used generically to cover a range of chemical reactions and processes that convert synthesis gas into synthetic crude oil, transport fuels, and various chemicals. (The synthesis gas is produced from natural gas as outlined in chapter 7.)

Literature. The literature on the FT process and related technology is vast. For instance, there are tens of thousands of patents. Extensive documentation was extracted from the major German and Japanese operating companies after World War II. This includes interrogation reports of the principal scientists and engineers. Many of these are available from the larger lending libraries on microfiche. A U.S. DOE project brought a considerable number of these documents to wider availability on the Internet.[2] In addition to this primary source data, the U.S. Bureau of Mines conducted extensive studies on the catalyst formulations. This work was extensively published and reviewed in major articles and available in books by R. B. Anderson.[3] Also of note are the reviews and book by M. E. Dry at Sasol, who gives valuable descriptions of the operations of large operating plants.[4]

Overview of the FT Processes

The direct conversion of synthesis gas into transport fuels has been practiced since the 1930s. In this process, named after the early German pioneers in this field, the conversion is brought about by catalysts of iron, cobalt, nickel, or ruthenium. The first two metals are of primary importance. Early plants were based on coal as a feedstock. The principal chemical reactions of the FT synthesis are:

Synthesis:

$$nCO + 2nH_2 = \text{-}[CH_2]_n\text{-} + nH_2O \qquad (10.1)$$

Methanation:

$$CO + 3H_2 = CH_4 + H_2O \qquad (10.2)$$

Water gas shift:

$$CO + H_2O = CO_2 + H_2 \qquad (10.3)$$

In the synthesis equation (10.1) $\text{-}[CH_2]_n\text{-}$ represents all of the hydrocarbon molecules from methane to paraffin wax of high molecular weight. Methanation represents methane made in the process over and above that produced by the synthesis reaction (10.1; n=1). The presence of the WGS reaction (10.3) during synthesis can be an advantage in helping to generate any extra hydrogen needed. Carbon dioxide is also a product when the WGS reaction is present.

The relative amounts of these reactions depend on the catalyst formulation and the temperature of the reaction. By judicious choice of catalyst, the FT process can span the extremes of producing hydrocarbons from entirely methane to entirely wax. The oxygen by-product can be entirely water or substantially carbon dioxide. Of note is that catalysts based on iron tend to have high WGS activity, and cobalt catalysts tend to have a high activity for methanation.

Relative to gas, coal has an excess of carbon, and as a consequence, the synthesis gas produced from coal has a low hydrogen to carbon monoxide ratio. The stoichiometric ratio is typically 0.5 to 1.0. Methane, made from coal—commonly referred to as synthetic natural gas (SNG)—can be a more useful fuel than coal itself. As a consequence, a certain amount of methanation (10.2) in the coal-based FT process can be advantageous.

The synthesis reaction and the methanation reactions are very exothermic. In addition, there is a large contraction in the volume of the gases as a consequence of the reactions. For methanation, the enthalpy change is –206 kJ/mol (298K). The volume contraction (the ratio of the product volume to reactant gas volume) is 50%. For the synthesis of cetane ($C_{16}H_{34}$), the enthalpy change is –2,473 kJ/mol, with a volume contraction of 66%.

A generalized layout for the production of a synthetic crude oil (syncrude) from natural gas is illustrated in figure 10–1.

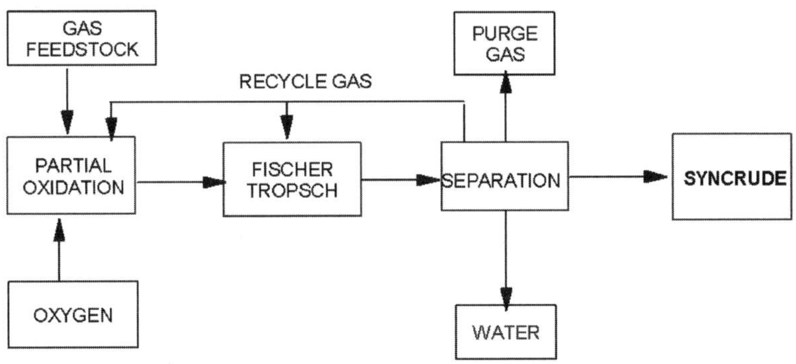

Fig. 10–1. Gas to synthetic crude oil

There are relatively few unit operations. Gas is converted into synthesis gas by partial oxidation, which is then passed to the FTR, where the crude products are formed. There then follows a separation section to separate the liquid products from the gas products and unreacted synthesis gas. The gas stream is further separated into a purge stream and a recycle stream.

With natural gas (methane) as the feedstock, carbon loss as carbon dioxide represents wasted natural gas. Methanation, since it represents formation of the starting material, adds to the processing costs. For these reasons, the older FT catalysts and processes developed for coal would be expected to be less than optimum for a process utilizing natural gas as the feedstock.

The synthesis gas required for the FT process may be produced by partial oxidation (as shown in fig. 10–1) or steam reforming. The latter gives a synthesis gas containing high levels of carbon dioxide. This carbon dioxide is only incorporated in the FT synthesis at high temperatures (in essence via the reverse of the WGS reaction). The synthesis gas produced using the former method would probably be hydrogen deficient relative to transport fuels. Thus a certain amount of hydrogen formation (WGS) in the synthesis would be advantageous.

There does not seem to be an obvious choice between a partial oxidation/iron catalyst system and a steam reforming/cobalt catalyst system, or any variation thereof. This leads to various approaches by purveyors of the technology.

The alpha value

The FT process produces a hydrocarbon product distribution that can be described by simple polymerization theory. The molar ratio of a product of a given carbon number (CN) to a product of one less carbon number is a constant less than unity. This constant is generally referred to as the *alpha value*.

$$[\text{Moles of product with CN} = n]/[\text{moles product of CN} = n - 1] = \alpha \qquad (10.4)$$

This can be modified to predict the mass fraction (W_n) of product with carbon number n, as:

$$\log (W_n/n) = n \log \alpha + \log [(1-\alpha^2)/\alpha] \qquad (10.5)$$

The alpha value is an important indicator of the nature of the products that are formed in the synthesis. The value of alpha depends on all of the independent variables in the synthesis. These include the temperature, pressure, stoichiometric ratio of the synthesis gas, the catalyst type, and the age of the catalyst.

The alpha value is a good predictor of the components produced in the range of C_5 to about C_{30}. This is the range of interest for the production of transport fuels. The amount of methane is more than that predicted depending on the extent of methanation. Ethane is particularly low, and propane and butane are reduced from the ideal. The amount of heavy materials can be reduced by cracking reactions.

The ideal variation of product composition is illustrated in figure 10–2.

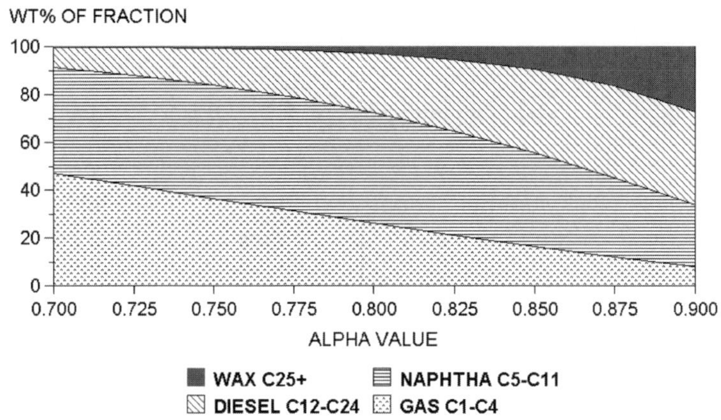

Fig. 10–2. Variation of product composition with alpha

Although some systems can operate at lower or higher alpha values, the typical FT processes of interest operate in an alpha range of 0.7 to 0.9. As the value of alpha rises, there is a fall in the production of gaseous products (methane to butane) and a concomitant rise in the production of wax (C_{25}+). Over much of the range of alpha, the liquid products (the sum of naphtha and distillate) remain essentially constant at about 55% to 60% of the total products. The principal problem for the production of naphtha and diesel is to maximize this liquid yield by secondary processing of the gas and wax fractions. This greatly complicates the overall process flowsheet.

The value of alpha can be altered by using catalysts of different formulations, and to some extent by changing process parameters, particularly temperature. (This is because increasing the temperature lowers the value of alpha.) However, in practice a specific technology usually operates over a restricted alpha range.

The product from the FT reaction can best be regarded as a synthetic crude oil. There is a very high (in some cases almost 100%) concentration of linear paraffins. Such paraffins make good kerosene and distillate blend stock, but the naphtha fraction is very poor in octane number.

Before use as gasoline, the naphtha must be extensively refined. Also produced are heavy waxes that can be hydrocracked to lighter transport fuel fractions. Thus unlike MTG (chapter 9), the products of FT require extensive treatment before use as transport fuels.

Maximizing Diesel

Because of its superior properties, many gas conversion strategies tend to dwell on the production of motor diesel (distillate) rather than gasoline. The demand to produce distillate or distillate-rich products from a synthetic fuels plant is seen as market driven rather than technical. As will be explained later, large volumes of distillate are not easy to produce.

Unfortunately, there seems to be no clear-cut definition of distillate in the realms of GTL process technology. To some extent, the distillation cut points in refineries can be altered to help cater to changes in demand for specific products. Figure 10–3 illustrates how typical refinery cut points are reflected in the composition of FT products.

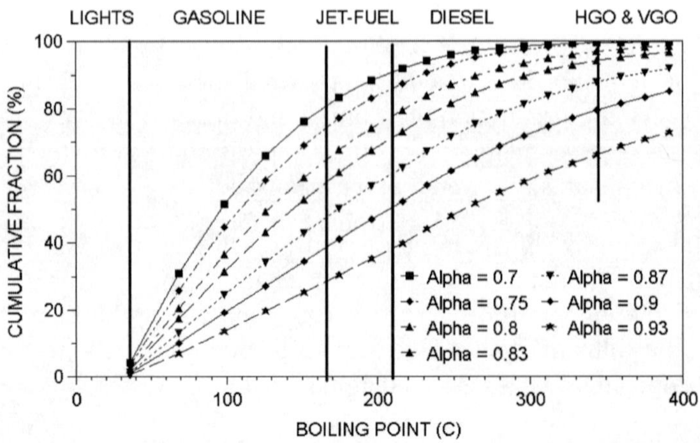

Fig. 10–3. FT product and refinery cut points

The figure plots the cumulative fraction against the boiling point of the nongaseous fraction of the FT products for a range of alpha values. Vertical bars indicate the typical refinery cut points, reflecting the common boiling ranges of the specification fuels listed.

However, organizations have their own product definitions. This manifests itself in patents and brochures describing synthetic fuels technology. For instance, Shell often defines distillate in its middle distillate synthesis (MDS) technology as 140°C–370°C. Mobil, in its MOGD technology, defines distillate as > 330°F (> 165°C). These differences in definition are important in relation to the usage. Distillate encompasses kerosene and a large portion of the heavy gasoline fraction (at the lighter end of the scale). It also includes the fuel oils (HGO and VGO) well into the other end of the spectrum.

This flexibility in the definition of distillate, in terms of boiling point, means that it is relatively easy to claim a 50/50 (or higher) distillate/gasoline split without further processing. Comparison of fuel specifications for motor diesel clearly indicates the anomaly.

Distillate via olefins or wax cracking

Straight-run distillate comprises a maximum of about 40% of the total product. However, more distillate can be made by either oligomerizing lighter olefins of the gas and naphtha fractions or by cracking the wax products in subsequent downstream processes. These are the bases of the Synthol (developed by Sasol and used at the Mossgas plant in South Africa) and the Shell MDS (used at Bintulu in Malaysia) processes, respectively.

The Synthol process operates at a relatively low alpha value (about 0.75) and at a high temperature (the latter helps maximize olefins). The paraffins are passed to a dehydrogenation unit to make more olefins. All the olefins are then oligomerized into diesel.

The Shell process operates at a high alpha value (> 0.9). This maximizes wax production. The product, which would be solid at room temperature, is then passed to a hydrocracker that produces diesel. These approaches are summarized in figure 10–4.

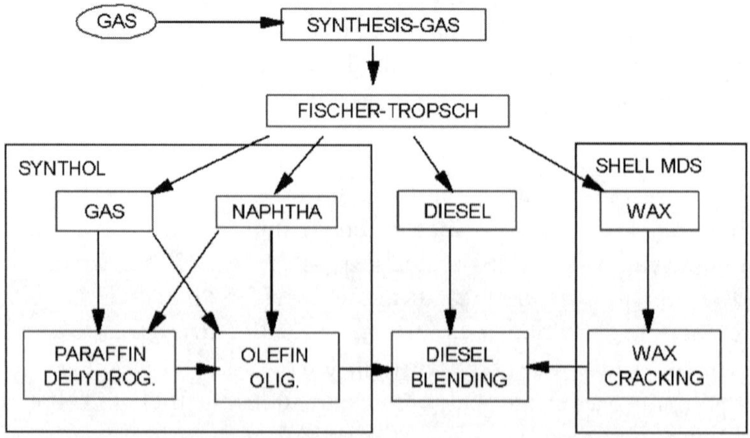

Fig. 10–4. Maximizing diesel

Processes

The principal issues for the successful practical use of the FT process are the removal of the large quantities of reaction heat and handling the concomitant volume contraction. The latter results in a lower volume of gas available to carry away the reaction heat. During World War II, scientists in German industry attacked this issue on a broad front. They developed, to near commercial scale, many of the reactor types known today for handling very exothermic reactions. These are:

- **Multitubular fixed bed (MTFB).** These reactors comprise a large number of small diameter tubes containing the catalyst. Boiling pressurized water to carry away the reaction heat surrounds the tubes.
- **High gas recycle/fixed bed.** A thin bed of catalyst is used with low pass conversion. After cooling and separation of the condensable products, large volumes of the unreacted gas are recycled (the recycle ratio is > 10).
- **Entrained bed.** The catalyst bed is moved around a loop by a large flow of gas. Sections of the loop are cooled to remove the reaction heat.

- **Fluidized bed.** The catalyst bed is fluidized by up-flowing reactant gas. Agitation of the catalyst with gas ensures effective heat transfer of the reaction heat from the catalyst.
- **Slurry bed.** The catalyst is held within a liquid slurry, which absorbs the reaction heat and transfers the heat to heat-exchange equipment within the reactor.

All except the high gas recycle option have been used as the basis of today's commercial FT processes.

German World War II technology

Table 10–1 gives the location and years of operation of FT plants controlled by Germany and its allies up to the end of World War II. Commercial production started in 1938. However, in Germany it effectively ceased in 1944 as a consequence of strategic bombing.

Table 10–1. Fischer-Tropsch plants of the Axis allies 1938–1945

Company	Location	Reactor type	Years of operation
Germany			
Ruhrchemie AG	Oberhausen-Holten	MTFB	1938–1944
Rheinpreussen	Moers-Meerbeck		1938–1944
Klockenerwerke	Castrop-Rauxel		1938–1944
Chemische Werke	Wanne-Eickel		1939–1944
Hoesch-Benzin	Dortmund		1939–1944
Braunkohle Benzin	Schwarzheide		1938–1944
Wintershall	Lutzendorf-Mucheln		1943–1944
Shaffgotsch Benzin	Odertal		1941–1944
France			
Couriers Kuhlman	Harnes	MTFB	1944
Japan			
Nippon Jinso Sekiyu	Miike Works Kyushu	MTFB	1940–1945
Nippon Jinso Sekiyu	Amagasaki, Honshu		1943–1945
Nippon Jinso Sekiyu	Takikawa, Hokkaido		1942–1945
Manshu Jinso Sekiyu	Chielin, Manchuria		
Manshu Gosei Nenryo	Chinchu, Manchuria		

All of the commercial operations in Germany, France, and Japan used MTFB reactors and used cobalt as the catalyst. The French and Japanese plants used technology supplied by Ruhrchemie. Some of these operations were major works, employing from about 40 to 300 reactors. All plants used coal as a primary feedstock, which was gasified to form synthesis gas in a variety of ways. There was also widespread development of iron catalysts and many trials of different reactor types using both iron and cobalt systems.

The main thrust of the German technology was to maximize the use of the very high quality straight-run diesel (cetane greater than 90). There was a major downstream industry utilizing various cuts to produce soaps, lubricating oils, and synthetic fats.

South African commercial operations

After the World War II, most major developments were conducted in South Africa. Table 10–2 summarizes the South African commercial operations.

Table 10–2. South African commercial operations

Company	Location	Reactor type	Years of operation
Sasol	Sasolburg (Sasol 1)	ARGE – MTFB	1954 to date
Sasol	Sasolburg (Sasol 1)	SYNTHOL (Entrained Bed)	1954 to date
Sasol	Secunda (Sasol 11)	SYNTHOL (Entrained Bed)	1980 to date
Sasol	Secunda (Sasol 111)	SYNTHOL (Entrained Bed)	1983 to date
Moss-Gas	Mossel Bay	SYNTHOL (Entrained Bed)	1991 to date

The ARGE process (Sasol). The last stage development of the MTFB system by Germany in World War II was an iron-based catalyst system with added recycle gas. A joint effort called in German Aitkengesellschaft (abbreviated to ARGE) of Lurgi and Ruhrchemie designed and built the first South African facility at Sasolburg in 1954.

By changing the operating variables and catalyst formulation, the ARGE process can operate at any level of alpha value. However, operating with alpha values greater than 0.8, it is the best-known method for the production of waxes and distillate. Early commercial activity used thermal cracking of wax to produce more distillate. However, today the process is optimized to produce wax, which is refined and sold as a specialty chemical.

The SYNTHOL process (Secunda and Mossgas). The primary focus of the process is the production of gasoline. Synthol is a development of a Kellogg entrained-bed process, which in turn was based on German developments during World War II. Entrained (and fluid-bed) technologies are best used for the production of light products. They operate with alpha values below about 0.75.

Synthol is the established method of producing most of the synthetic fuels at Sasolburg, Secunda, and Mossgas in South Africa. The operation involves the production of an olefin-rich product in very large entrained-bed reactors. In the Mossgas plant, synthesis gas is produced from natural gas by a combination of partial oxidation and steam reforming. The other plants use coal as the primary feedstock. The products are easily separated from the catalyst and are processed to result in both gasoline and diesel fuel in a large number of subsequent process operations.

Key features of the process are:

- Synthol generates a large quantity of methane. This can be beneficial with coal as the feedstock. However, if natural gas is used, it has to be returned to the natural gas oxidation plant or the reformer.
- Synthol produces a significant amount of product in the range of 160°C–350°C that is good quality distillate, as witnessed by its high cetane number of 55.
- Gasoline quality is poor (octane is about 60), and extensive refining is required.

Another feature of the Synthol process is the relatively large quantity of water-soluble products (acids, esters, and ethers). These require extraction before final water disposal.

Overall, the Synthol route balances a poor innate carbon efficiency (a large volume of light gases is produced) against a relatively rich olefin stream product. The olefin stream product is comprised of 86% olefins in C_3 and C_4, which are 24% of total products. These are supplemented by significant quantities of good quality kerosene and motor distillate. Further details of the product distribution are given in chapter 9.

Other commercial operations

At the time of writing, there are several proposals to use the FT process in major facilities, but there are only two other commercial operations of note.

Shell MDS. The Shell MDS plant at Bintulu in Malaysia commenced operation in 1995. This is a MTFB process basically very similar to ARGE, with the exception of a cobalt catalyst. The process is also similar to a technology developed by Gulf, which experimented with cobalt and ruthenium catalyst combinations.

Natural gas is turned into synthesis gas by partial oxidation (using Shell technology). The FT process produces a heavy wax, which is hydrocracked into middle distillates. The plant has the capacity to remove and refine waxes and linear paraffins.

Standard Oil. Standard oil operated a very large scale demonstration (semicommercial) project at Brownsville in Texas between 1955 and 1959. The plant was gas based, with partial oxidation as the method for producing synthesis gas, and it employed a large fluid-bed reactor. The principal product was gasoline, and both iron and cobalt catalysts were used.

Processes under development

Over the years, many large and small corporations have worked on the development of the FT process. Some notable ones active at the present time are presented here.

Slurry reactors. Although MTFB reactors are used to produce heavier distillate products, these reactors are very large, and the productivity (output) is relatively low. The entrained-bed system has higher productivity but only gives lighter products. Several additional operations are needed to produce distillate.

The slurry-bed system has been known since the 1940s and was extensively researched by Kobel during the 1950s.[5] It offers the potential of producing heavier distillates and wax at considerably higher productivity (hence smaller size, lower weight, and lower cost) than the MTFB system.[6] The basis of the slurry reactor is shown in figure 10–5.

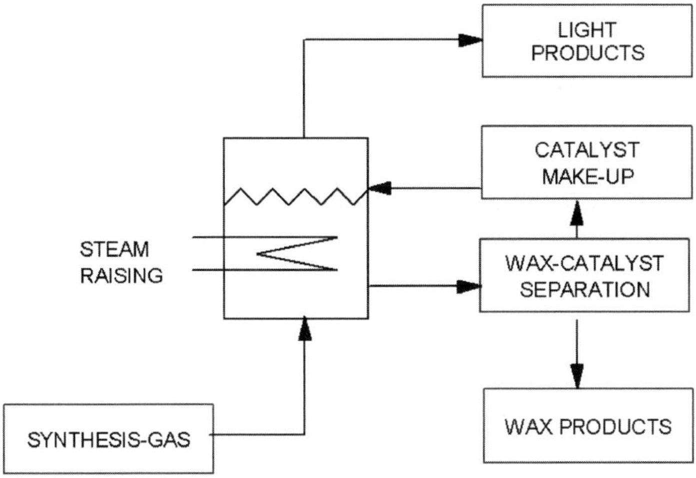

Fig. 10–5. Slurry FT reactor

In the slurry reactor, synthesis gas enters the bottom of a slurry of molten wax and catalyst. Reaction heat is efficiently transferred to the slurry and then to steam-raising coils within the slurry. Light products boiling below the operational temperature (typically 200°C for wax and distillate products) distill out of the slurry. Heavier material (distillate and wax) collects in the slurry. Part of the slurry is drawn off, and the wax and catalysts are separated. The catalyst is returned to the reactor. Despite its simplicity, there are several complications that have delayed its introduction as an alternative to the MTFB. Some issues are:

- Catalyst attrition in the slurry results in fine-particle blockages in the filters used to separate the distillate/wax from the catalyst to be recycled.
- As the operation proceeds, the nature of the slurry changes, and this can result in the deposition or plating out of the catalyst.
- Carbon monoxide and hydrogen have different solubilities in the slurry. These have to be accounted for in the synthesis gas feed.

As well as Kobel's work, the slurry system has been extensively researched by the major companies in the field, including Sasol, Shell, and Exxon-Mobil. The system has also been researched by technology development companies such as Syntroleum and Rentech. Sasol in particular has been in the forefront of developments, with both high- and low-temperature variants in operation.[7] Sasol has three 1,500-bbl/d semicommercial plants operating at Secunda. At the time of writing, some of the major GTL plants proposed for the Middle East and Nigeria have entered the engineering and construction phases using slurry reactors.

The BP-Arco process. The BP-Arco process is being demonstrated on a 300-bbl/d pilot unit at Nikiski in Alaska. This technology concentrates on a novel approach to the production of synthesis gas by a compact reformer developed in conjunction with Davy Process Technology. The proponents of the process claim that the reformer is easier to modularize and has a lower overall weight and cost than conventional reformers.[8]

Sasol Chevron. Sasol and Chevron have combined forces in a joint venture to promote the FT-GTL technology.[9] The joint venture uses Sasol's slurry-bed process to produce a waxy primary product that is then cracked using Chevron's established isocracking process.[10] Sasol Chevron is providing the technology for a 33,000-bbl/d plant at Escravos in Nigeria.[11]

Exxon AGC-21. Exxon has been researching FT processes since World War II, and as Standard Oil operated the previously mentioned large semicommercial plant at Brownsville. Their latest developments (AGC-21) concentrate on using a novel catalytic partial oxidation/ steam reforming unit. This unit employs a fluidized-bed system to tailor the synthesis gas stoichiometric ratio to the requirements of a slurry-bed reactor.

Syntroleum. The Syntroleum system uses air-based autothermal reforming to produce a synthesis gas diluted with nitrogen. This is used in a cobalt-catalyzed slurry reactor to produce the product.[12] This approach (like the BP approach) eliminates the need for an oxygen plant. Syntroleum is promoting this concept for offshore (floating) plants.[13]

Other companies. Rentech has built several small-scale plants aimed at utilizing low-value gas streams, such as landfill gas, to produce high-valued wax products. Their approach is generally to use steam reforming to produce synthesis gas, followed by an iron-based slurry reactor. Conoco has built a pilot plant in order to evaluate its CoPOX process, a catalytic process for producing synthesis gas and its proprietary FT technology.[14] The U.S. DOE is funding research into novel membrane reactors that eliminate the need for air separation plants with the intention of coupling to FT reactors. These are discussed in chapter 7.

Products Made by the FT Process

The FT process produces a range of normal paraffins. These can be converted into commercially salable products as explained here.

Transport fuels

The primary product has the properties of a crude oil with the benefit of negligible sulfur, nitrogen, and oxygen content. In addition, the middle distillate fraction has very good properties as assayed by standard refinery methods. As such it is an excellent feed or co-feed for refinery operations. Table 10–3 gives the properties of a typical FT syncrude produced on a MTFB reactor operating at an alpha value of 0.84.

Table 10–3. Assay of FT syncrude

Property	Method	Units	Value
Gravity	ASTM D1298	Deg. API	56.7
Sulfur	IP 336	Wt.%	0.004
Pour Point	ASTM D97	Deg. C	42
Acid Number	ASTM D664	mg KOH/gm	1.45
Viscosity	ASTM D445	cSt @ 80C	1.13

Operating at an alpha value of 0.84, the syncrude has a high API gravity (low density). Sulfur, as measured by IP Method 336, is present in a concentration of 40 ppm. However, this is probably a consequence of the imprecision of this method for determining low sulfur content in fuels rather than the actual sulfur content. In fact, other methods indicate that the sulfur content of the crude is extremely low (< 10 ppm). [The correct methods for the analysis of low sulfur content fuels are a matter of current debate. R. A. Kishmore Nadkharni of Millennium

Analytics Inc. (American Laboratory, November 2000), has described a European study on the precision of the various methods for determining sulfur in fuels, with particular emphasis on 50 ppm gasoline and diesel. This study indicates a large error in method IP 336 (which is equivalent to ASTM D 4294: EDXRF), and it is not recommended for determining the sulfur content at low levels. The levels quoted in these tables probably represent maximum values.]

The syncrude, being rich in linear paraffins, is very waxy and has a high pour point (> 42°C). In other words, it is solid at room temperature.

The high acid number (> 1.4) is indicative of the presence of some acids (carboxylic acids) formed as by-products in the process. High synthesis temperatures tend to favor more oxygenates, such as alcohols and carboxylic acids, in the product.

Although solid at room temperature, once melted, the syncrude is a free-flowing, low-viscosity clear liquid.

Transport fuel blend stocks and specification transport fuels are produced as described in the following section.

Gasoline. The straight-run gasoline fraction of the FT product described above has the properties described in table 10–4.

Table 10–4. Properties of FT gasoline

Property	Method	Units	Value
Boiling range		Deg. C	23.7–190.6
Yield on crude	ASTM D86	Vol. %	43
Gravity	ASTM D1298	Deg. API	70.1
Specific gravity	ASTM D1298	60/60 F	0.701
Sulfur	IP 336	ppm	50
RON	ASTM D2699	Clear	20.9

Working with an alpha value of 0.84, the straight-run gasoline yield is 43% of the syncrude. Sulfur is very low. Note the commentary concerning the applicability of the test method. It has a very low research octane number (RON). This can be improved by isomerization and reforming the naphtha fraction. Straight-run gasoline produced by methods operating at a higher temperature and lower alpha value (e.g., Synthol) has higher octane. This is a consequence of higher olefin content (RON typically 50).

Additional gasoline can be produced by oligomerization of light olefins (propylene, butene, and pentene). This is practiced in the South African Synthol operations. Cracking heavier fractions can also make gasoline. Vapor phase cracking that gave a reasonable octane level (> 85) was practiced during World War II.

Motor diesel fuel. The properties of the straight-run distillate fractions of the previously discussed FT syncrude are given in table 10–5.

Table 10–5. Properties of FT straight run distillates

Property	Method	Units	Value	Value
Boiling Range	ASTM D86	Deg. C	190.6–232.3	232.3–315.8
Yield on Crude	ASTM D86	Vol. %	11.4	18.3
Gravity	ASTM D1298	Deg. API	56.6	50.2
Specific Gravity	ASTM 1298	60/60F	0.752	0.778
Sulfur	IP 336	ppm	50	40
Smoke Point	IP 57	mm	>50	
Aniline Point	ASTM D611	Deg. C	79.2	90.6
Diesel Index	IP 21		80.7	98

The properties of two distillate fractions are given. The lower b.p. range (196°C–232°C) corresponds to a kerosene (typical jet fuel) b.p. range. The yield on the syncrude is about 11%. The higher boiling range (232°C–316°C) corresponds to a motor diesel fraction. The yield on the syncrude is 18%.

Relative to conventional refinery streams, these two products have high API gravity (low density). Indeed, an important point is that the heavier fraction corresponding to motor diesel fuel has a gravity below the normal specification of motor diesel fuel (minimum density of 0.820).

Again the fractions are very low in sulfur. FT diesel is usually observed to have a sulfur content below 10 ppm when measured by a method appropriate for determining very low sulfur values.

Straight-run diesel has a very high cetane as represented by the diesel index, typically over 70. This high cetane value and low sulfur content is the basis for the attraction of FT diesel. However, note that the density of the diesel is very low. In fact, it is lower than the acceptable level in the new clean fuel standards being introduced. This makes the FT diesel more of interest as a blendstock for higher density (nonspecification) fuels than as a fuel in its own right.

More diesel can be produced by oligomerizing olefins in the naphtha fraction (Synthol) and cracking the heavier gas oil fractions (ARGE and Shell MDS). Because oligomerization and cracking introduce more branched molecules, the diesels produced from these processes are inferior to the straight-run diesels in their cetane values.

Chemicals

It is true that the use of the FT synthesis for the production of chemicals is not a primary aim. However, the process does produce a range of intermediates that could be separated and used in a variety of downstream applications. These include:

- **Alcohols and derivatives.** There are a wide variety of processes that fall under the banner of FT processes that produce alcohols as the major products. These can be short-chain alcohols, particularly butanols, as well as longer chain alcohols. Butanol synthesis has been used in the past.[15]
- **Olefins.** Operating with a low hydrogen concentration in the synthesis gas promotes the formation of alpha olefins. Using low-temperature synthesis such as ARGE gives higher olefins with a high degree of linearity. Such olefins are sought after for linear alcohol production.
- **Linear paraffins.** The products from the ARGE type of process are paraffins of high linearity. These can be extracted and used as solvents or chemical feedstocks (e.g., linear alkylbenzenes, LAB).
- **Waxes.** The ARGE process can produce paraffin waxes of very high linearity. These materials are highly marketable. Both the ARGE operations in South Africa and the Shell operations in Bintulu extract and refine wax products.

- **Lubricating oils.** Extensive isomerization of the heavier diesel and wax fractions produces high molecular weight, highly branched paraffins, with negligible sulfur content. These have beneficial properties as lubricant basestocks (so-called Group 3 basestock). Some proposed GTL schemes have targeted this as a major product.

Economics

Unless otherwise stated, the products of the GTL FT process are assumed to be naphtha and diesel as separate products. Most technologies concentrate on the production of diesel or a mode that will maximize the diesel yield. For ease of analysis, these products are taken together as a mix. Syncrude comprises an unseparated mixture of naphtha, diesel, and wax. Data is developed for stand-alone plants with the location factors applied as necessary for a U.S. Gulf Coast operation.

Commercial operations

For the most part, economic analysis can only be performed after consulting several sources and using appropriate judgement to estimate some of the key factors.

The Shell MDS plant at Bintulu in Malaysia was described in the early part of its operation in several articles published by Shell.[16] Since commencing operation, the plant has undergone modification. This led to the production of large volumes of wax (to be sold as product) and the reduction of bottlenecks to increase plant capacity to about 14,000 bbl/d.

There is a paucity of information about the Mossgas plant, which uses the Synthol process. Data have been gleaned from several sources.[17] The Sasol Chevron joint venture has produced estimates for an idealized FT operation and has defined a planned target in terms of capital and gas usage.[18] Capital cost breakdown of a typical Sasol-type operation has been published by workers at Foster Wheeler Ltd. and is illustrated in figure 10–6.[19]

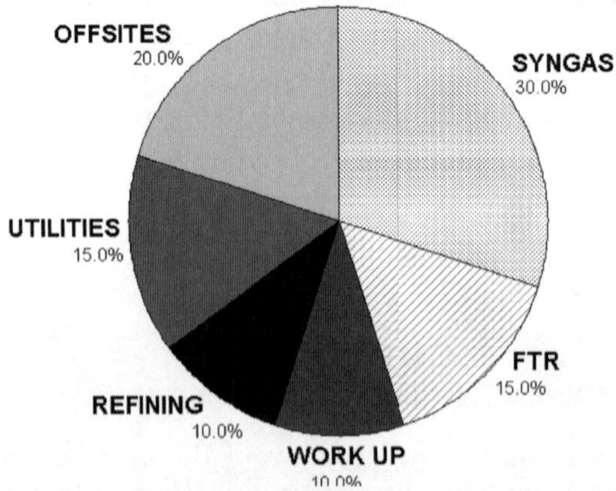

Fig 10–6. FT capital breakdown. From B. Ghaemmaghami and S. C. Clarke. "Study Yields Generic, Coastal-Based GTL Plant," *Oil & Gas Journal*, March 12, 2001, p. 64.

Note the dominance of the cost of producing synthesis gas in the capital breakdown. The reactor section (FTR) only comprises 15% of the total capital.

The corresponding cash-flow breakdowns are illustrated in figure 10–7. These cash flows are similar to those estimated for the work presented here.

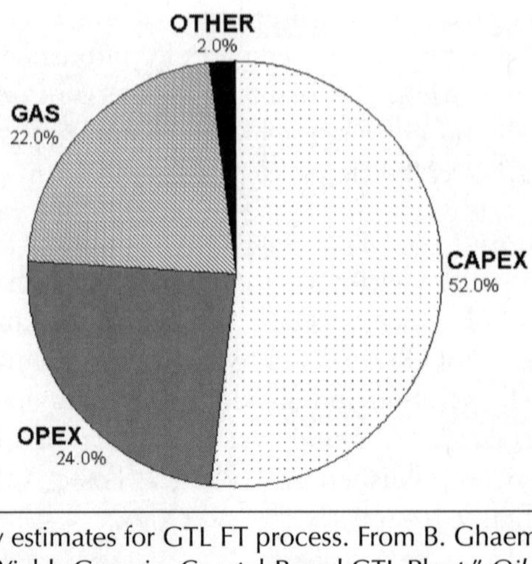

Fig. 10–7. Cash-flow estimates for GTL FT process. From B. Ghaemmaghami and S. C. Clarke. "Study Yields Generic, Coastal-Based GTL Plant," *Oil & Gas Journal*, March 12, 2001, p. 64.

The base data for developing fixed-variable equations shown in figure 10–8 are given in table 10–6. The data have been escalated to a 2004 cost base.

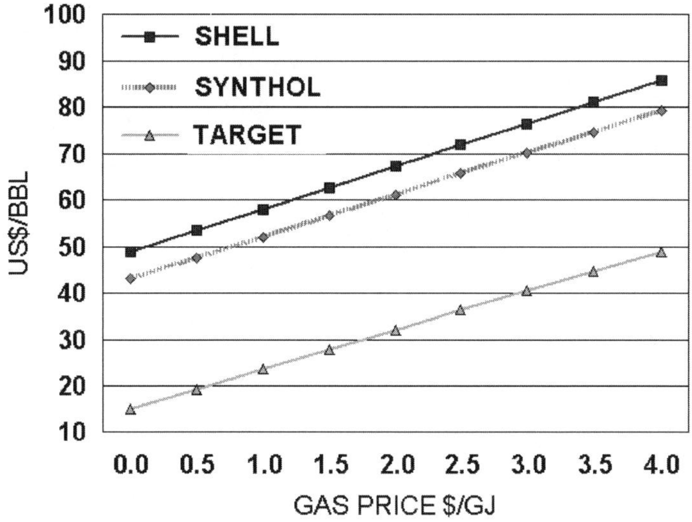

Fig. 10–8. FT unit production costs

Table 10–6. Statistics for Fischer-Tropsch GTL

		Synthol	Shell	Target
Production	kt/a	692.24	470.00	610.61
	bbl/d	17005	11332	15000
Capital costs	$MM	934.5	842.9	360.9
Construction period	years	3	3	3
Plant life	years	15	15	15
Return on capital	%/a	16.34%	16.34%	16.34%
Non gas operating costs	$MM/a	249.08	188.29	76.98
Gas usage	PJ/a	52.3	35.6	43.2

Three cases are presented. The Shell MDS case is based on the original plant design, which has now been extensively reworked to eliminate bottlenecks. The Synthol case is similar to the operation at Mossgas. Within the errors associated with these types of concept estimates, both cases give a similar outcome. For gas at $1/GJ, both produce product in the $50/bbl to $60/bbl range. The breakdown of capital and operating costs is similar to that reported by workers at Foster Wheeler.

The third case is a target detailed in the Foster Wheeler work. This target requires significant capital reduction and lower operating costs, as well as improved gas-conversion efficiency over the other two cases. For gas at $1/GJ, this ideal plant would produce product below $30 bbl.

For comparison, the variation in the cost of naphtha, gasoline, and gas oil (which is the primary feedstock for diesel) with crude oil (Tapis) on the Singapore market is given in figure 10–9. Further details of discussion of the cases are given below.

Mossgas. The South African government promoted the use of the gas field at Mossel Bay for synthetic fuels. The technology chosen is an FT process in the variant known as Synthol. In this process, synthesis gas is converted at relatively high temperatures using a fluid-bed catalyst to olefins and a naphtha-rich synthetic fuel. Synthesis gas is produced by a combination of steam reforming and partial oxidation. Next the product gases are mixed to obtain the required hydrogen to carbon monoxide ratio.

Undoubtedly the prime motivation for choosing this route is that the South Africans have complete control of the technology. It is based on their considerable experience with the conversion of coal to fuels at Secunda. In the South African context of the Apartheid Era, the chosen technology was secure.

The emphasis in South Africa is to produce large quantities of olefins as intermediates that are oligomerized into distillate range fuels. Traditional polymerization technology is difficult and produces a highly branched (hence poor quality) product. Thus IFP's Polynaphtha technology is used.

This choice of technology, bringing with it the inefficiencies of steam reforming and the FT process, is unlikely to be economically viable in countries with more liberal economies. The development is very large, costing in the order of US$850 million (R2,000 million), and producing in excess of 1 MMt/y refined fuels, of which 50% is distillate. The fixed-variable relationship indicates that at $1/GJ, the production cost would be about $50/bbl.

Shell MDS technology. The project is capable of producing about 12,000 bbl/day of product. For commercial viability, the product value would have to be greater than about $50/bbl using gas priced at about $1/GJ.

In order to improve the commercial viability of the project, Shell is maximizing product value by extracting and refining high-value paraffins and waxes. Shell is also maximizing the residual product value by blending it with inferior low-value stocks and producing specification fuels.

Sasol Chevron target. The Foster-Wheeler publications (see fig. 10–6 and 10–7) give an indication of what might be expected in future FT plants. The target is for a 15,000-bbl/d plant that would cost an estimated $330 million. This would produce product at $30/bbl with gas below $2/GJ. On the face of this analysis, such a project would be viable. It is important to note that this target requires a better performance than is expected from the Nigeria operation under development.

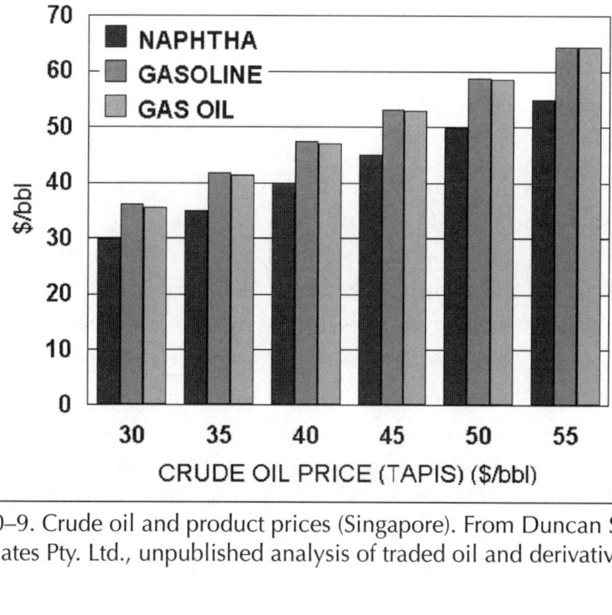

Fig. 10–9. Crude oil and product prices (Singapore). From Duncan Seddon & Associates Pty. Ltd., unpublished analysis of traded oil and derivative prices.

Processes under development

Robertson has presented an economic analysis of some of the FT technologies under development.[20] This work was part of an analysis of GTL options for Alaska's undeveloped gas fields of the North Slope, which are estimated to contain 38 Tcf of gas. The complexities in this analysis (such as location factors and location options) make a direct comparison with the analysis presented here difficult. The salient points of the analysis are given in figure 10–10. This compares the capital charges and operating charges for several variants.

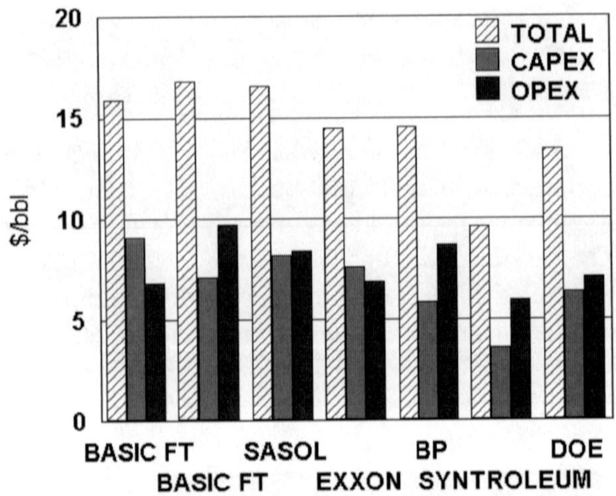

Fig. 10–10. Comparison of capital costs for various GTL technologies

The operating charges are high as a result of the extreme environment. The results shown in the figure indicate that there is little difference between the various options studied apart from the Syntroleum technology. This offers significantly lower costs as a consequence of the use of air and hence nitrogen-diluted synthesis gas. (However, it is the author's opinion that the analysis does not take full account of the extraction costs from FT product streams diluted with nitrogen from the use of an air-based process. In the author's experience, the consequence of high nitrogen content gas is to require downstream vessels to be considerably larger. The products are more difficult to fully extract. This increases the overall cost so that little, if any, cost saving results.)

One aspect of the studies into the use of GTL on Alaska's North Slope is the question of transporting the products to market. Crude oil from the Prudhoe Bay oilfields is moved by pipeline to Valdez. The crude oil is heavy (24° to 32° API), so the question arises as to the suitability of using the pipeline to move the much-lighter and waxy GTL products.[21]

Reengineered World War II technology

In a series of papers, Antia and Seddon have proposed the reengineering of technology from the World War II era and also public domain technology for niche applications.[22] This work concentrated on small-scale plants using about 4 PJ/y gas (10 MMcf/d) to produce a syncrude that would be blended into the produced oil. In some variants, broad range diesel would be produced and used immediately to offset diesel imported onto a remote site or FPSO.

Pertinent data are given in table 10–7, and the fixed variable equation for two such schemes are illustrated in figure 10–11.

Two cases are presented. The first is for a small plant producing syncrude. The capital costs are on an ISBL basis, and the plant is intended to be integrated with a FPSO or similar operation. The small-scale option has a similar fixed-variable relationship to the commercial plants illustrated in figure 10–8. The second is for a larger stand-alone operation consuming about 30 PJ/y gas. The larger scale reengineered technology has a similar economic outcome to that proposed for the optimum FT process (target) discussed previously.

Table 10–7. Statistics for re-engineered WWII technology

		ISBL	Optimum
Production	kt/a	35.1	573.9
	bbl/d	854.2	14098.0
Capital costs	$MM	39.9	233.1
Construction period	years	2	3
plant life	years	15	15
Return on capital	%/a	15.49%	16.34%
Non gas operating costs	$MM/a	9.8	58.1
Gas usage	PJ/a	4.1	35.1

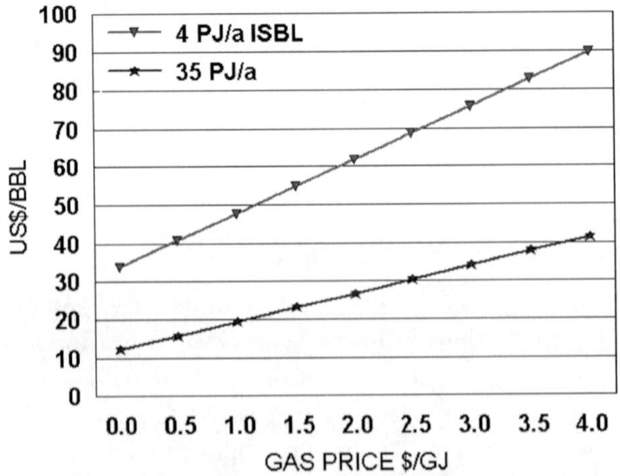

Fig. 10–11. Reengineered World War II FT unit production costs

Conclusion

Although there has been a great deal of interest and activity in the FT process, commercial development has been piecemeal and slow. The technology suffers from large capital costs relative to conventional fuels. In order to be commercially competitive, this GTL route requires gas to be available for less than $2/GJ and also requires concomitant high oil prices.

The principal interest is the production of diesel fuel, as this is of very high quality and contains no sulfur. Nowadays, the most common approach involves the use of slurry reactors to produce a waxy product that is then hydrocracked to form diesel. This approach gives the highest carbon efficiency (typically about 75%). However, the overall energy efficiency remains low (typically 60% to 65%).

Despite a range of new developments and approaches, cost improvements remain marginal and there is no outstanding technology. Many of the newer approaches differentiate themselves on the method for the production of synthesis gas. None of these approaches (including the use of slurry reactors) has been technically proven on the scales envisaged for commercial operation. A critique of the FT GTL route by Bakhtiari came to similar conclusions.[23]

References

1. "Worldwide Construction," *Oil & Gas Journal,* November 25, 2002, and April 25 2005.
2. U.S. Department of Energy. "German Document Retrieval Project," sponsored by Syntroleum and developed by A. Stranges at Texas A&M University, available at www.fischer-tropsch.org.
3. Anderson, R. B. "Catalysts for the Fischer-Tropsch Synthesis," *Catalysis*, vol. IV, P. H. Emmett, editor, New York: Reinhold Publication Corp., 1956.
4. Dry, M. E. "The Fischer-Tropsch Synthesis," *Catalysis Science & Technology*, J. R. Anderson and M. Boudart, editors, vol. 1, Springer-Verlag, 1981.
5. Kolbel, H., and M. Ralek. "The Fischer-Tropsch Synthesis in the Liquid Phase," *Catalysis Reviews—Science and Engineering,* 21(2), 1980, p. 225.
6. Singleton, A. H. "Advances Make Gas-to-Liquids Process Competitive for Remote Locations," *Oil & Gas Journal*, August 4, 1997, p. 68.
7. Chang, T. "South African Company Commercializes New F-T Process," *Oil & Gas Journal*, January 10, 2000.
8. Font Freide, J., T. Gamlin, and M. Ashley. "The Ultimate 'Clean' Fuel—Gas-To-Liquid Products," *Hydrocarbon Processing*, February 2003, p. 53; and "Gas Processes 2002," *Hydrocarbon Processing*, May 2002, p. 105.
9. Chang, T., "New JV Markets One-Stop GTL package," *Oil & Gas Journal*, December 18, 2000, p. 46.
10. "Refining 2000," *Hydrocarbon Processing*, November 2000, p. 116.
11. "Escravos Project Adds GTL to Phase 3; Nears Phase 2 Start up," *Oil & Gas Journal*, October 23, 2000, p. 62.
12. "Gas Processes 2002," May 2002, p. 106; U.S. Patent 6,265,453.
13. Bigger, J. M., and H. L. Tomlinson. "Consider Barge-Mounted Plant to Produce Ultra Clean Diesel," *Hydrocarbon Processing*, July 2004, p. 39.
14. "Gas Processes 2002," May 2002, p. 105.
15. Cohn, E. M. "The IsoSynthesis," *Catalysis,* vol. IV, P. H. Emmett, editor, New York: Reinhold Publication Corp., 1956.

16. van der Burgt, M., J. van Klinken, and T. Sie. "The Shell Middle Distillate Synthesis Process," selected papers, Shell Ltd, November 1989; and G. A. Bekker. "A First for Shell in Bintulu," *Oil & Gas News*, October 1990.

17. Technical descriptions of Sasol technology are found in the publications of Dry, such as M. E. Dry. "High Yield Quality Diesel from Fischer-Tropsch Process," *Chem SA,* February 1984, p. 286.

18. Chang, T. "New JV Markets One-Stop GTL Package," December 18, 2000, p. 46.

19. Ghaemmaghami, B., and S. C. Clarke. "Study Yields Generic, Coastal-Based GTL Plant," *Oil & Gas Journal*, March 12, 2001, p. 64.

20. Robertson, E. P., INEEL (Bechtel), available on the Web site of Alaska Department of Revenue and summarized in "North Slope Alaska GTL Options Analyzed," *Oil & Gas Journal*, January 31, 2000, p. 74.

21. Khataniar, S., et al. "Technical and Economic Issues in Transportation of GTL Products from Alaskan North Slope to the Markets," Society of Petroleum Engineers Technical paper SPE 86931, abridged as, "Transportation of GTL products from Alaskan North Slope," *Journal of Petroleum Technology*, April 2004, p. 56.

22. Antia, D. D. J., and D. Seddon. "Low Cost 10MMcf/d Gas to Syncrude Plant for Associated Gas," Offshore Technology Conference, Houston Texas, May 1998, OTC Paper 8901; ibid. "Gas Conversion to Syncrude," World Expo, 1996, pp. 87–96; ibid. "Offshore Conversion of Associated Gas to Synthetic Crude Oil: An Economic Option for Deep Water and Marginal Fields." Offshore Technology Conference, Houston Texas, May 1995, OTC Paper 7868; ibid. "Improving the Economics of Developing Very Deep Water Fields through the Conversion of Associated Gas to Crude Oil," Deeptech95, February 28–March 2, 1995, Aberdeen; ibid. "Gas Conversion: An Economic Alternative to Gas Reinjection," Offshore South East Asia (OSEA94) 10th Conference and Exhibition, Singapore, December 6–9, 1994, Paper 94003; ibid., "Offshore Refining: A Cost Effective Approach for Treating Associated Gas," European Petroleum Conference, October 25–27, London, United Kingdom, EUROPEC 94, 1994, SPE Paper 28858; and "Economics of Gas Conversion Projects in the North Sea and Barents Sea," EUROPEC90, The Haag, Netherlands, October 22–23, 1990, SPE Paper 20937.

22. Bakhtiari, A. M. "Gas-to-Liquids: Much Smoke, Little Fire," *Hydrocarbon Processing*, December 2001, p. 20.

11

Liquefied and Compressed Natural Gas— LNG and CNG

This chapter discusses liquefied and compressed natural gas. LNG is a globally traded commodity. After transcontinental pipelines, it is the most common method used to transport gas over long distances from remote fields to markets. It is the method of choice if the construction of a pipeline is impossible for some reason.

LNG operations are very large and usually involve joint ventures (JV). They comprise four parts, which are often separated into different companies. The JV partners will have different shareholdings depending on their specific interests. The four parts typically are:

1. Large-scale development of a gas field, delivering typically 1 PJ/d (1 Bcf/d) of gas
2. Gas treatment and liquefaction to produce LNG and storage
3. Shipping of LNG in specialized transoceanic vessels
4. Storage, regasification, and use of LNG

This chapter will concentrate on the liquefaction of LNG with some reference to its shipping, storage, and regasification. Because LNG production and shipping is expensive, compressed natural gas (CNG) has been proposed as an alternative. Because this technology is similar to LNG, it is also discussed in this chapter.

LNG Production Methods

The production of LNG is a physical separation process. Unlike chemical processes such as methanol production or FT GTL, there are no major opportunities to improve the fundamentals of the process. As a consequence, the engineering design is aimed at obtaining as close a match as possible to the ideal thermodynamic behavior for the separation. There are several different engineering approaches to this that differentiate the proprietary technologies. Meeting the thermodynamic limit is costly in terms of the capital equipment required, and therefore maximizing the scale of operation minimizes production costs. A typical world-scale LNG plant would have a capacity to process more than 1 Bcf/d (300 PJ/y).

From the economic standpoint, small-scale LNG schemes suffer adversely from trying to meet the thermodynamic limit. Therefore cheaper plants, working away from the ideal, are used. The downside is a loss in efficiency, which can be measured as the quantity of feed gas required to produce a unit amount of LNG.

Both large- and small-scale operations follow the same strategy, which involves gas cleaning followed by progressive cooling of the feed gas until liquefaction occurs.

Gas pretreatment

For LNG projects, it is critically important to have clean gas feedstock comprised only of methane and ethane. Trace impurities, especially mercury, are detrimental to the process operation. Typically the gas feed to an LNG production unit will have the specifications given in table 11–1.

The typical unit operations required to produce such a feed are illustrated in figure 11–1.

Table 11–1. Specifications for the gas feed to LNG

Nitrogen	vol%	< 1%
Carbon dioxide	ppm (vol)	< 50
Hydrogen sulphide	ppm (vol)	< 4
Total Sulfur (H_2S + COS + mercaptans)	ppm (vol)	< 20
Water	ppm (vol)	< 0.5
Mercury	micro-gm/cm	< 0.01
Butanes (max)	vol%	2%
Pentanes (max)	vol%	0.10%
BTX (max)	vol%	Not detected

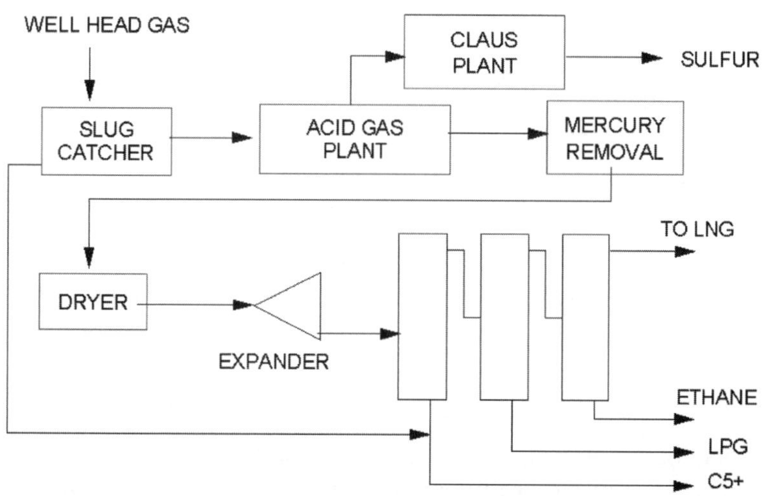

Fig. 11–1. Gas pretreatment for LNG

The figure is illustrative of the number of unit operations involved.

- The raw gas is passed to a slug catcher, where condensate is removed.
- An acid gas plant removes carbon dioxide, hydrogen sulfide, and other sulfur compounds. The off gas is passed to a Claus plant to recover the sulfur.
- Mercury removal is essential to protect the downstream liquefaction train.
- The gases are dried to remove moisture and other trace impurities.
- Turbo expansion cools the gas, and then condensate, LPG, and (optionally) ethane are progressively removed.

In the scale of operation, the extraction of the condensable materials will produce significant quantities of by-product streams of LPG and condensate. These products make a major contribution to the economics of the overall LNG operations.

Liquefaction

There are three general approaches to liquefaction technology.[1] These are the cascade process, the mixed refrigerant process, and the expander cycle process.

The cascade process. The general layout of the cascade liquefaction process is illustrated in figure 11–2. The approach to the thermodynamic limit is achieved by means of a series of cold boxes operating with different refrigerants. Propane, ethylene, and methane are the most commonly used refrigerant gases. Best practice is achieved by operating these separate refrigerants at different pressures (and hence temperatures) within their respective cold boxes and by close integration of the cold boxes. This is achieved in the Phillips Optimized Cascade LNG Process. After liquefaction, the LNG is passed to large cryogenic holding tanks. The cascade process is the most efficient process, but capital costs are high, and hence it is best suited to large train sizes.

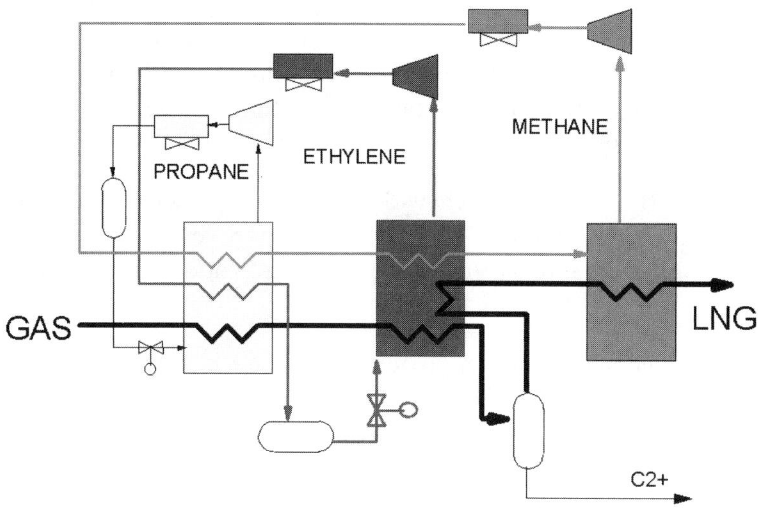

Fig. 11–2. Cascade process

Mixed refrigerant process. This is the most commonly used process. In this approach, the three refrigerants are mixed and operated over a range of pressures and temperatures. This process is illustrated in figure 11–3 for the single-flow mixed refrigerant process.

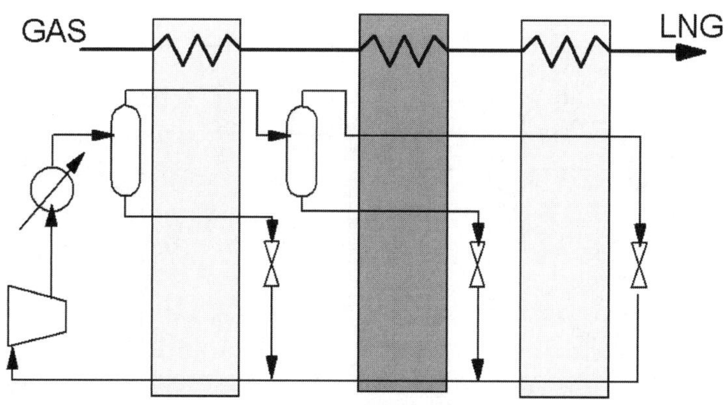

Fig. 11–3. Single-flow mixed refrigerant process

Mixed refrigerant processes (MRP) can have higher thermodynamic efficiencies than the cascade process, but overall efficiency is lower because higher refrigerant flows are required. However, the MRP has a simple configuration, and capital costs are therefore lower than those of the cascade system.

The MRP comes in several variants. The single MRP is commonly used for the smaller peak-shaving schemes.[2] These schemes are used by gas utilities to store gas in a distribution network in order to service peak demand loads.

For larger plants, one approach often used is an initial cold box operating with propane like the cascade process. This is followed by a cold box operating with mixed refrigerant. Another variation (AP-X APCI Process) has the mixed refrigerant cold box followed by a cold box using nitrogen as the refrigerant in an expander process.

Expander cycle process. This is simply the use of a single refrigerant (usually nitrogen or methane) working with a compression and expansion turbine delivering the cooling. Expander processes are easy to start up and shut down and are often favored for peak shaving processes. However, the process has relatively high power consumption. Systems with two expanders are common and are used for the liquefaction of air (fig. 11–4).

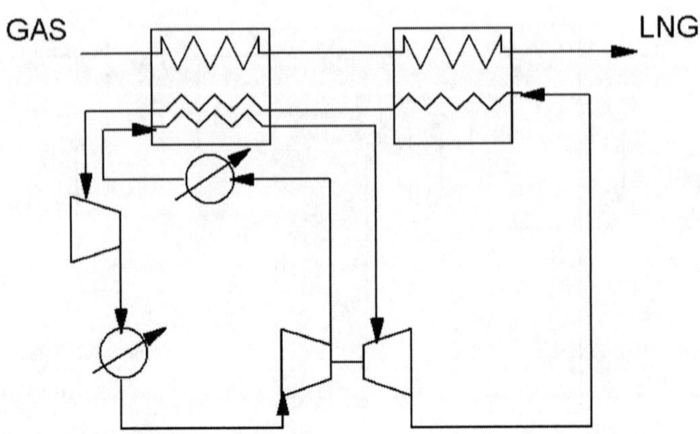

Fig. 11–4. Expander process

Thermal efficiency. The actual efficiencies achieved by large-scale plants can be about 90% (LNG energy out/energy in). However, taking into account all site operations (gas plant, etc.), actual site efficiency is about 85%.

Smaller scale operations that are not confined to operating near the thermodynamic limit employ fewer stages. In this case the efficiency is lower. It seems likely that smaller plants would mainly operate in the vicinity of 75% overall efficiency.

Floating production of LNG (FLNG)

The use of LNG plants to process associated gas or gas from smaller fields is being extensively studied. The technology choice is governed by the constraints of the FPSO. The preferred liquefaction technology appears to be based on a dual expander refrigeration cycle.[3] Foglietta has described a typical process.[4] This type of process offers the following advantages:

- Process simplicity
- Reduced plot area
- Modularization
- Safe and simple operation
- Minimal flare requirements
- No refrigerant inventory
- No motion impact
- Equipment reliability

A typical process is illustrated in figure 11–5.

In this process three expansions are employed to cool the gas to liquid. There are two independent refrigeration cycles. The methane cycle cools both the gas and the nitrogen for the nitrogen cycle. The nitrogen cycle further reduces the gas temperature to near the liquefaction point. An expander on the gas stream then produces the LNG.

In the methane cycle, methane is compressed to about 7 MPa (1,000 psia), cooled to about –7°C (20°F), and then expanded to about 1.4 MPa (200 psia). This cools the methane to –84°C (–120°F). The gas reenters the cold box and exits having been warmed to near ambient temperature in order to recommence the cycle.

In the nitrogen cycle, high-pressure nitrogen at 8 MPa (1,200 psia) at near ambient temperature enters the cold box and is cooled to –90°C (–130°F). The gas exits the box and is expanded to about 1.4 MPa (200 psia), cooling to –162°C (–260°F). The gas reenters the cold box and warms back to near ambient temperature in order to recommence the cycle by recompression to 8 MPa.

High-pressure gas at about 7 MPa (1,000 psia) enters the cold box and is cooled by the refrigeration gases to about –160°C (–255°F). The gas leaves the cold box and is expanded to about 150 kPa (20 psia), liquefying the gas.

For the production of 0.5 MMt/y LNG (75 MMscfd), approximately 263 MMscfd methane and 172 MMscfd nitrogen are used in the refrigeration cycles.

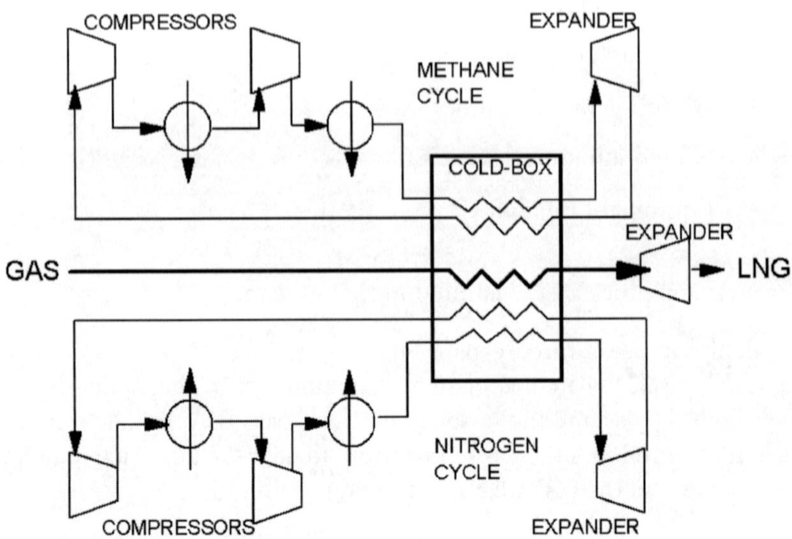

Fig. 11–5. Dual expander process for FLNG

These units can be produced in modules with the dimensions given in table 11–2.

Table 11–2. Dimensions for LNG modules

Capacity MMt/a	Length m (ft)	Width m (ft)	Weight (tons)
0.5	64 (210)	20 (65)	1,450
1	64 (210)	37 (120)	2,180
1.5	64 (210)	56 (185)	3,260

In addition to the fractionation plant, there are other key issues to be addressed in floating LNG operations. One particular issue is the design of towers for fractionation or absorption. Packed columns are the preferred approach in offshore floating (moving) environments.[5]

Market for LNG

Since its inception in 1970, the market for LNG has grown enormously. The growth in the world market is illustrated in figure 11–6.

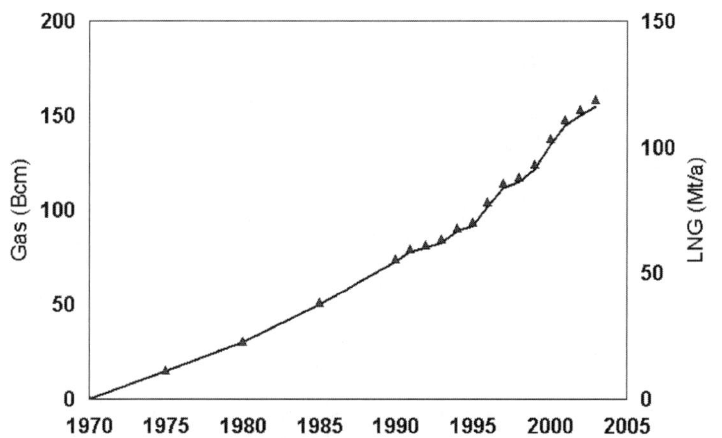

Fig. 11–6. LNG trade. From C. T. Sen. "World's LNG Industry Surges, Pushed by Confluence of Factors," *Oil & Gas Journal*, June 14, 2004, p. 58.

This rapid growth rate, which currently represents an increase of about 7 MMt/y, has resulted in the growth of LNG projects. The train sizes have grown from 1 MMT t/y to projects comprising multiple trains of 3 MMt/y each.

The main importing and exporting countries in 2003 are shown in tables 11–3 and 11–4, respectively.

Table 11–3. LNG Importers (2003)

	Mt/a	Bcm/a	% total
Japan	58.51	72.17	46.7%
Korea	19.41	23.88	15.5%
Spain	12.11	15.08	9.7%
U.S.	10.54	13.32	8.4%
Puerto Rico	0.54	0.68	0.4%
France	9.21	11.51	7.4%
Taiwan	5.59	6.9	4.5%
Turkey	3.45	4.31	2.8%
Belgium	2.52	3.16	2.0%
Italy	2.49	3.12	2.0%
Greece	0.42	0.53	0.3%
Dom. Rep.	0.24	0.31	0.2%
Portugal	0.18	0.22	0.1%
Total	**125.21**	**155.19**	**100.0%**

Table 11–4. LNG exporters (2003)

	Mt/a	Bcm/a	% total
Indonesia	26.54	32.85	21.2%
Algeria	21.29	26.70	17.0%
Malaysia	16.84	20.71	13.5%
Qatar	14.31	17.56	11.4%
Trinidad	8.84	11.23	7.1%
Nigeria	8.74	10.85	7.0%
Australia	7.64	9.42	6.1%
Brunei	7.10	8.71	5.7%
Oman	6.79	8.33	5.4%
Abu Dhabi	5.26	6.49	4.2%
U.S.	1.29	1.65	1.0%
Libya	0.56	0.68	0.4%
Total	125.20	155.18	100.0%

LNG composition

All complexes produce LNG with a methane content of about 90%. However, there are differences in the concentrations of the various heavier hydrocarbons present.

In many cases, LNG importers are part of a specific LNG project. They help specify the LNG product composition.

This may prove a problem if LNG becomes widely available from alternative suppliers.[6] It is not clear whether LNG from different suppliers with different compositions is interchangeable. One problem is the potential for heavier liquid fallout as the LNG is regasified.[7]

Yang and others have discussed the issue of removal of heavier material in order to produce a consistent product for the U.S. market.[8] Huang and others have compared compression methods to condensing methods for removing ethane and higher components.[9]

Production Costs

The approach is to use as a basis all plant and associated costs typical of a major land-based LNG installation for a U.S. Gulf Coast site. From this figure, variations for other locations can easily be factored into the total costs.

Of the capital cost, gas liquefaction accounts for less than 50% of the total. The rest typically is made up of LNG storage (18%), utilities (16%), loading facilities (10%), and gas pretreatment (6%).[10]

The fixed-variable relationship is shown in figure 11–7. This represents the base case for the large-scale production (6 MMt) of LNG. The pertinent statistics are given in table 11–5.

The capital cost has been estimated from the reported costs for recent plants.[11] It has then been adjusted for compatibility with a U.S. Gulf Coast site. A scaling factor of 0.65 has been applied.[12] This is to allow for improved plant efficiency in terms of capital costs per tonne of LNG produced, due to increased plant sizes. There is still considerable interest in improving efficiencies.[13] Some groups have proposed the use of all-electric drives instead of turbines in order to improve throughput.[14]

The overall complex thermal efficiency was estimated at about 85%. Avidan and Martinez have estimated the efficiency for the refrigerant section in various designs to be between 91% and 93%.[15] However, this is reduced here to reflect the costs needed to produce ancillary services and to permit an operating allowance, etc.

By-product credits are for LPG extracted in the gas treatment section of the plant. The rate is calculated from Woodside Petroleum's operation in Australia at 87,000 t LPG/MMt LNG.[16] Condensate, which can have a major positive impact on total project economics, was regarded as being part of the wellhead gas operation. By-product credits have been ignored for condensate.

Figure 11–7 shows that for wellhead gas at $1/GJ, the base case production cost of LNG is $222/t, or $4.09/GJ.

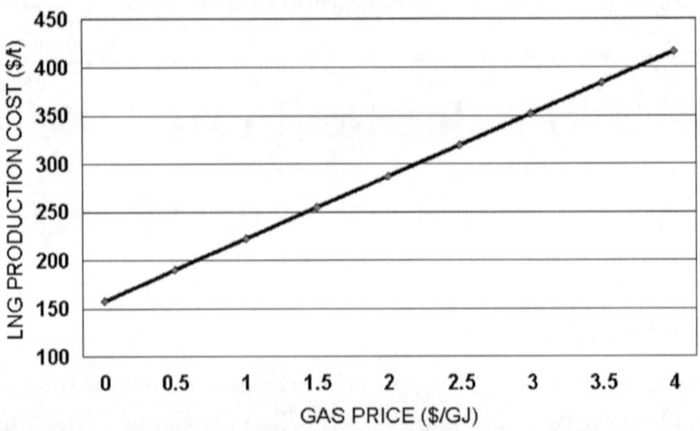

Fig. 11–7. LNG production cost

Table 11–5. Statistics for the production of LNG

Production	kt/a	6000
Capital cost	$MM	5225.5
Construction period	years	3
Plant life	years	20
Return on capital	%/a	14.60%
Non gas operating cost	$MM/a	1018.7
Gas usage	PJ/a	388
By-product credits	$MM/a	70.52

Transport of LNG

Transport costs are also discussed in chapter 5. Unlike other energy cargoes, the cost of transporting LNG is high in energy terms. Most vessels are dedicated and owned either by gas suppliers or the utilities receiving the gas. In other words, there is no competitive vessel chartering, as is the case for shipping crude oil or other oil derivatives.[17]

The cost of vessels is high, typically $200 million each. A world-scale LNG project may require the dedicated use of six or more vessels. Vessel costs can thus be $1.2 billion or more for each LNG project. Transport logistics is a major feature of LNG projects and a major cost item.

Typical transport costs are given in figure 11–8 (pertinent statistics are given in chapter 5).[18]

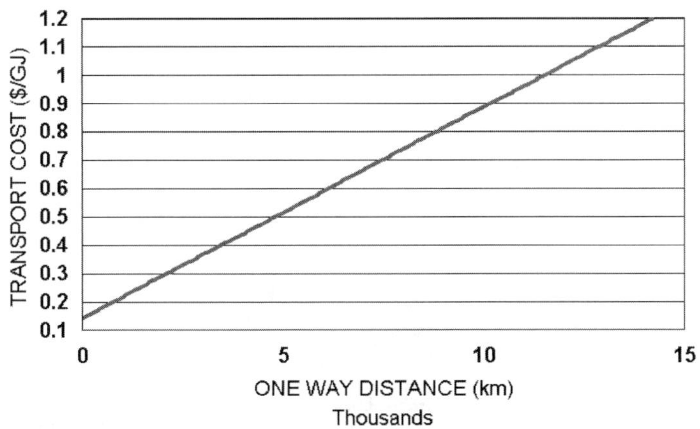

Fig. 11–8. LNG transport cost

The figure plots the cost of LNG transport in terms of dollars per gigajoule against the one-way distance of the voyage. For a typical 5,000-km voyage, the cost of LNG transport is about $0.51/GJ.

Boil-off

One issue of concern with shipping is the boil-off that occurs during both loading and transport. The boil-off during the voyage is nowadays handled by using the boil-off instead of diesel to power the ship. The amount of boil-off is dependent on the length of the journey but is typically about 2.5% to 3% of the cargo. This includes the boil-off on the return journey, when approximately 5% of the cargo is left in the vessel in order to provide ballast and maintain the holding tanks at a cryogenic temperature.

The extent of boil-off during loading can be problematic, as the LNG evaporates when it first enters a relatively warm storage vessel. If the boil-off can be returned to the LNG process, then this is done. Otherwise it is often vented or flared, which represents a considerable financial loss, as well having an environmental impact. The amount of boil-off has been estimated at a rate of 100 MMscfd over a 12-hour loading period.[19]

LNG Terminals

LNG is delivered at –164°C. In order to be of use, it has to be regasified to ambient temperature. After docking, the LNG ship discharges the liquid cargo to storage tanks from which the LNG is drawn to the regasification units.

Land storage

The LNG industry tends to follow European standards and definitions concerning storage tanks.[20] The tanks are large cryogenic vessels made of 9% nickel content steel. These tanks are insulated at the sides and base by a thick layer of pearlite or similar insulation material. A suspended aluminium deck covers the tank. This tank is then enclosed within a carbon steel tank for added protection. The whole arrangement rests on a concrete foundation.

The principal issue of concern is spillage and emission of LNG should the tank and its containment vessel rupture.

There are three types of tank containment: single containment, double containment, and full containment.[21]

Single containment. For a single-containment tank (fig. 11–9), a bund wall surrounds the tank. The wall is of a suitable height and distance from the tank so that the volume within its walls is greater that the contents of the tank. This is the standard type of containment in the petrochemical and refining industry. However, for LNG, should the liquid leak out, then the vapors from the (now boiling) LNG would be emitted directly to the atmosphere.

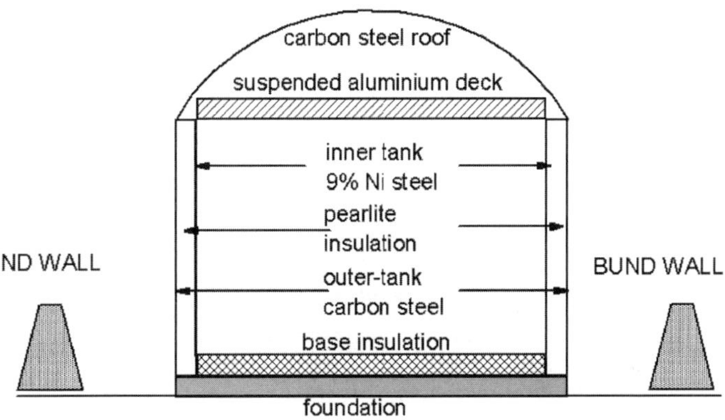

Fig. 11–9. Single-containment LNG tank

Double containment. In the double-containment tank (fig. 11–10), the bund wall is replaced by a prestressed concrete outer wall. This is very close to the tank, and the foundations are extended to meet this outer wall. Should the inner tank rupture, the liquid contents are contained within the outer wall, as is the case with the bund. However, the smaller volume and surface available to the spilled liquid would be expected to reduce the rate of vapor emissions. Nevertheless, should rupture occur, vapors will still be emitted to the atmosphere.

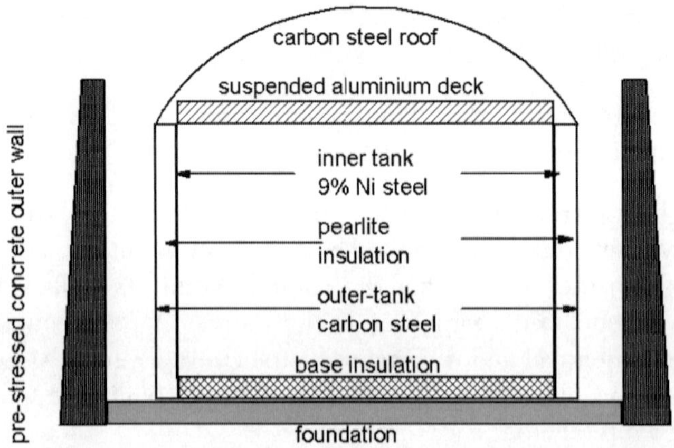

Fig. 11–10. Double-containment LNG tank

Full containment. The most common type of tank is the full-containment tank (fig. 11–11), where the concrete walls and roof form the outer tank. Should rupture of the inner tank occur, the LNG and the vapors are contained within the wall and the roof.

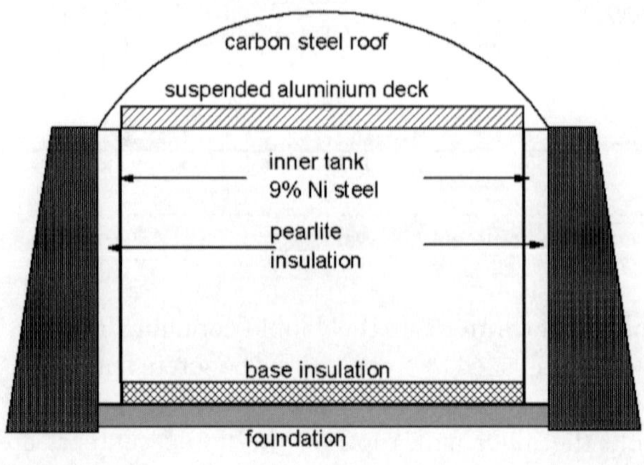

Fig. 11–11. Full-containment tank

Many regulatory authorities have eliminated the use of single-containment tanks by requiring a prestressed concrete outer wall. Many regulations enforce strict siting criteria that effectively eliminate the double-containment tank (e.g., EN1473). Other regulations in the United States (e.g., US 49CFR193) require the outer tank to be at least 10% larger than the inner tank. This prevents the use of full-containment tanks, and many double-containment tanks do not meet this criterion, either. As a consequence, many of the LNG tanks used in the United States are single-containment types.

While the regulatory regime for determining the type of tank remains confused, the use of underground caverns for the storage of regasified LNG has been proposed.[22]

Regasification

Regasification is accomplished by one of three methods: open rack vaporizer, submerged combustion vaporizer, or intermediate fluid vaporizer.[23] An open rack vaporizer is illustrated in figure 11–12.

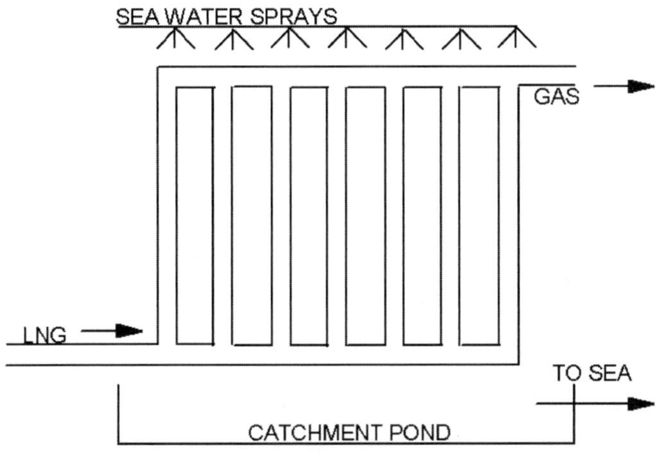

Fig. 11–12. LNG open rack vaporizer

This is the simplest type of vaporizer. LNG is passed through a series of pipes made from zinc-coated aluminium, over which seawater is sprayed. Heat is transferred from the seawater to the LNG, which becomes gas. The seawater is chilled in the process by 4°C to 5°C. Environmental considerations limit the seawater temperature drop by about 10°C before return. This means the seawater flow is very large, requiring extensive pumping.

The submerged combustion vaporizer (fig. 11–13) has a much smaller space requirement and does not require seawater.

In this type of vaporizer, LNG is passed via stainless steel pipes through a water bath heated by a gas burner system that is located under the water level. Thus maximum heat is transferred to the water. Exhaust gases from the burner exit the top of the water bath container. This type of system consumes some of the vaporized LNG as fuel. As well as having a small size, this type of vaporizer is capable of handling fluctuating loads.

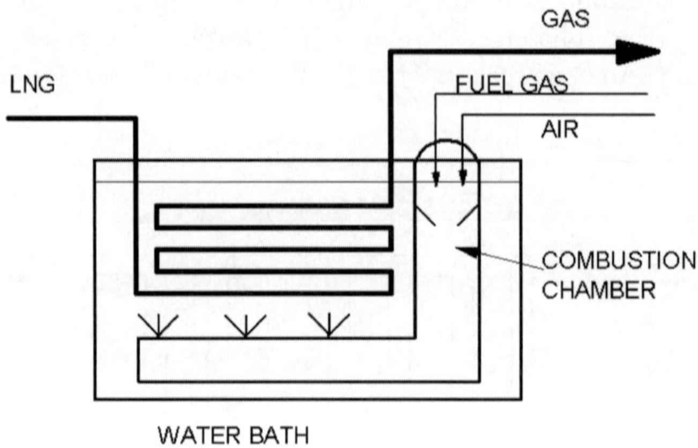

Fig. 11–13. LNG submerged combustion vaporizer

In the intermediate fluid vaporizer, LNG is gasified by means of a working fluid, which is in turn heated by seawater (fig. 11–14).

LNG enters a series of heat exchanger pipes that are warmed by a condensing refrigerant—propane, butane, or a similar refrigerant gas. The cold vaporized gas is passed to a natural gas heater, which brings the gas to ambient temperature. The natural gas heater and the refrigerant are warmed by a seawater circuit.

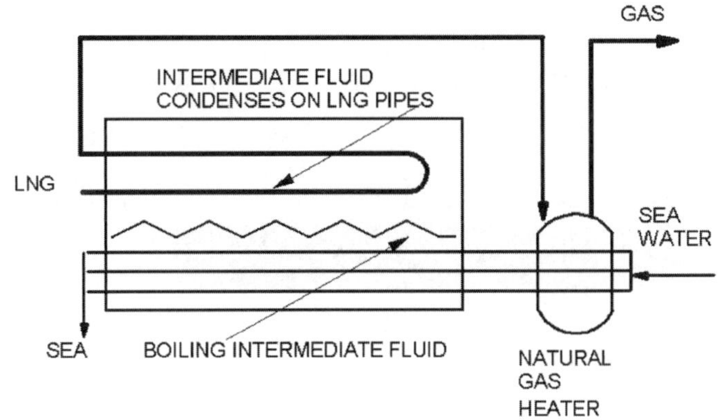

Fig. 11–14. LNG intermediate fluid vaporizer

Cold utilization

Warming the LNG with seawater or gas heaters throws away a valuable asset in the refrigeration power of a large mass of material at –164°C. This could be used beneficially in the support of other industries.[24] Japanese terminals best exemplify some of these uses.

Liquefaction and separation of air. Compression and expansion (the Joule-Thomson cycle) to the liquefaction point (–183°C) conventionally liquefies air. This air can then be separated to liquid nitrogen, oxygen, and optionally argon by distillation at cryogenic temperatures. The cost is dominated by the power of compression. This cost can be considerably offset by heat exchanging the air with LNG vaporization. This process is practiced at several terminals in Japan (table 11–6).

Table 11–6. Air liquefaction at some Japanese LNG terminals

Terminal	LNG used (t/h)	Liquid Oxygen (cm/h)	Liquid Nitrogen (cm/h)	Liquid Argon (cm/h)
Negishi	8	7000	3050	–
Senboku–1	23	7500	7500	150
Sodegaura	34	6000	9000	100
Chita	26	6000	4000	120
Senboku–11	40	7500	7500	200
Tobata	15	3500	3500	75
Niigata		3000	4000	–

Liquefaction of carbon dioxide. The required temperature to liquefy carbon dioxide (at pressure) is –55°C. LNG can be used to achieve this temperature by heat exchange. Some examples are given in table 11–7.

Table 11–7. Carbon dioxide production at some Japanese LNG terminals

Terminal	LNG used (t/h)	Liquid Carbon Dioxide (t/d)	Dry Ice (t/d)
Senboku –1	3.2	100	-
Chita	6	70	40
Negishi	6.4	86	48

Cryogenic power generation. As LNG is vaporized, it expands. Approximately 1 m^3 of LNG results in 600 m^3 of gas. This expansion can be utilized directly by natural gas expansion, NGE. It can also be indirectly utilized through an intermediate fluid using a Rankine cycle (R). The systems are quite complex and usually require a high electricity tariff to justify the investment. Some examples are given in table 11–8.

Table 11–8. Power generation at some Japanese LNG terminals

Terminal	Process	LNG used (t/h)	KW
Senboku –11	NGE + R	210	7,450
Himeji	NGE + R		3,000
Chita	NGE + R	300	15,400
Tobata	NGE + R	150	9,400
Niigata	NGE	175	5,600
Negishi	R	100	4,000
Higashi	NGE	270	11,300

Other schemes. There are several other methods of utilizing the LNG's low temperatures. For example, liquid nitrogen produced at the Senboku-II terminal is used in the manufacture of frozen food.

Regasification costs

The cost of storage and regasification is significant. In 2003 typical costs were about $400 million for a large import terminal.[25] Using this capital cost as the basis, the regasification cost can be estimated at $0.48/GJ (table 11–9).

Table 11–9. Estimate of LNG regasification costs

Nominal capacity	kt/a	3300
Capital cost (2004)	$MM	434.07
Construction period	y	2
Operating period	y	20
Capital recovery (10% DCF, royalty free)	%	13.57%
Return on capital	$MM/a	85.98
Operating costs	$MM/a	21.27
Gas output	PJ/a	179.51
Regasification cost	$/GJ	0.48

Others have estimated the regasification cost at typically about $0.30/MMBTU.[26]

Traded Prices

A relative few and closely allied companies carry the trade in LNG. Open discussion of traded LNG prices is rare. Unlike other energy products, the spot market for LNG is still in its infancy.

LNG price is generally linked to a competing fuel. In the United States, the competing fuel is pipeline natural gas. The price may be taken as either the long-term contract price or the Henry Hub spot gas price. In Europe, low-sulfur residual fuel is often used as the marker for LNG, and pipeline gas from Russia is also used as a reference. In the Pacific, where trade is dominated by LNG sales to Japan and to a lesser extent Korea, the cost is linked to the prevailing price of a crude oil derivative, typically gas oil.

When discussed, prices tend to be quoted cif Japan. Typically, prices during the mid-1990s were in the range of \$4.00/MMBTU to \$4.50/ MMBTU (GJ). From this, a quoted price of \$4.20/GJ cif Japan would mean a production cost in the region of \$3.40/GJ. This is in line with the estimated cost of production from \$1/GJ gas.

As the prices of oil and gas have risen dramatically, they have had a major impact on the landed price of LNG, and existing LNG operations are currently very profitable.

Compressed Natural Gas

The potential of using CNG as an alternative means of getting gas to market is receiving attention. A CNG system would have the advantages of:

- Elimination of the expensive cryogenic plant for the production of LNG
- Lower gas processing costs, as the exhaustive elimination of condensable components would not be required
- No requirement for regasification facilities at the receiving port
- Lower energy use to produce the CNG (estimated at about one-half that required for LNG).

The key is the development of new lighter storage methods that lower the cost of shipping.

Methods of containing CNG in ships

Early methods of containing CNG in pressure vessels aboard ships have proved too expensive to be viable. Three new methods are underdevelopment.

Coselle. This comprises a long (10-mile.), small diameter (6-in.) pipe coiled around a carousel. The capacity of the coselle is about 3.2 MMcf. The cost of the coselle is a fraction of the cost of a pressure vessel of similar capacity. A standard bulk carrier vessel is used to contain stacks of coselles, typically 108 coselles for a 60,000-DWT vessel. This would transport about 345 MMscf gas.

VOTRANS. This system uses large diameter pipes in an insulated cold storage unit. Carrying the gas at low temperatures allows lower pressures to be used than would be required for ambient storage.

GTM. The GTM system uses composite reinforced pressure vessels made from a high-strength, low-alloy (HSLA) steel pipe, wrapped with composites. GTMs are approximately 35% lighter than conventional steel tanks.

Economics

There are no examples of a CNG operation at this time. The cost of shipping would be the major component of a CNG project, which would typically be 80% to 85% of the total project cost. The shipping economics have been estimated from the data provided by Wagner.[27] This is illustrated in figure 11–15 using similar methodology to that developed for the LNG case. Pertinent statistics are given in table 11–10.

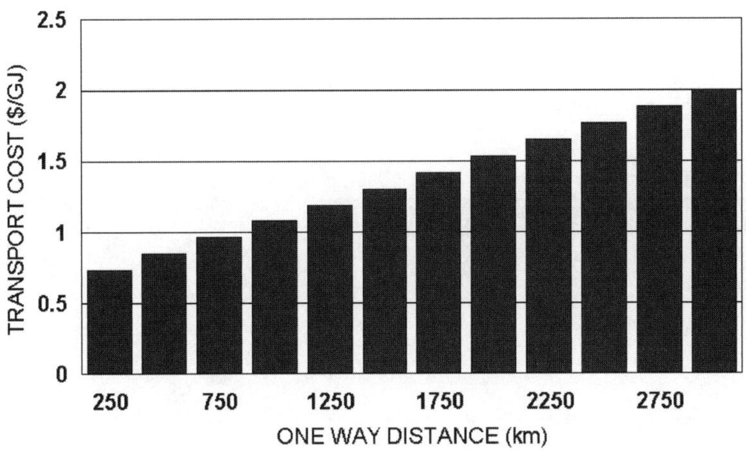

Fig. 11–15. CNG transport costs

Table 11–10. Statistics for a CNG carrier

Ship	DWT	60,000
Cargo weight	tonnes	7476.6
Speed	knots	18
Cost	MM$	110
Life	years	10
ROC (10% DCF)		15.19%
Port Fees	$/station	60,000
Maintenance, Fuel, Labor	MM$/a	7.66
Other Costs	MM$/a	2.16

CNG costs considerably more to ship than LNG. CNG would be of interest for relatively short distances (500 km to 3,000 km). At greater distances, the high cost of transport would likely make the gas uneconomic.

Wagner and Cone describe the use of CNG as an option for gas being available to an FPSO.[28] They compare the CNG option against the alternatives of a pipeline or an onboard LNG plant. The CNG facilities and the pipeline option are considerably lighter than the LNG facilities.

The case developed is for the production of 400 MMscfd of raw gas. After liquids removal, this produces 350 MMscfd of delivered gas by CNG or pipeline, and 314 MMscfd by LNG. From their analysis, Wagner and Cone show that the various options gave the payback (Capex/net revenue) detailed in table 11–11.

Table 11–11. Comparison payback for CNG, pipeline and LNG

	500km	1500km	3000km
CNG	0.60	0.70	1.30
Pipeline	1.20	2.70	5.30
LNG	3.00	3.30	4.00

Data from J. Wagner and S. Cone. *Oil & Gas Journal*, December 2, 2002, p. 68.

Greenhouse Impact

The greenhouse impact of LNG plants arises from the fugitive emissions of methane, which has a greenhouse impact of 21 times that of carbon dioxide. It also arises from the emission of carbon dioxide. The emission of carbon dioxide has a major impact, and this can arise from two main sources:

- The gas treatment plant when acid gases are removed
- The combustion of fuel in gas turbines to drive the refrigeration compressors

Past practice was to discharge carbon dioxide to the atmosphere from the gas plant. Now there are several LNG projects that propose the recovery and sequestration of the gas in order to ameliorate the greenhouse impact. Snohvit in Norway and Gorgon in Australia are two such projects. Both have available sequestration sites and both utilize a natural gas particularly high in carbon dioxide content.

However, for most cases of low carbon dioxide content gas (< 2% by volume), the carbon dioxide discharged from the gas plant is small compared to the emissions resulting from the combustion of gas in the turbines. For a typical operation working at an overall efficiency in the region of 85%, this amounts to about 0.5 t of carbon dioxide per tonne of LNG.

References

1. Finn, A. J., G. L. Johnson, and T. R. Tomlinson. "Developments in Natural Gas Liquefaction," *Hydrocarbon Processing*, April 1999, p. 47.

2. Price, B. C. "Small-Scale LNG Facility Development," *Hydrocarbon Processing*, January 2003, p. 37.

3. Finn, A. J. "New FPSO Design Produces LNG from Offshore Sources," *Oil & Gas Journal*, August 26, 2002, p. 56, and references therein; "BHP Unveils New Compact LNG Technology," *Oil & Gas Journal*, June 8, 1998; W. R. True. "Reducing Scale, increasing Flexibility Are Targets of New LNG Designs," *Oil & Gas Journal*, December 6, 1999, p. 54.

4. Foglietta, J. H. "Consider Dual Independent Expander Refrigeration for LNG Production," *Hydrocarbon Processing*, January 2004, p. 39.

5. Harrold, D. "Design of Turnkey Floating LNG Facility," *Hydrocarbon Processing*, July 2004, p. 47.

6. Bramoulle, Y., P. Morin, and J-Y. Capelle. "Differing Market Quality Specs Challenge LNG Producers," *Oil & Gas Journal*, October 11, 2004, p. 48.

7. Fatica, A. "Gas-Quality Debate Heats up As More U.S. LNG Imports Loom," *Oil & Gas Journal*, June 14, 2004, p. 69.

8. Yang, C. C., A. Kaplan, and Z. Huang. "Cost-Effective Design Reduces C_2 and C_3 at LNG Receiving Terminals," *Oil & Gas Journal*, May 26, 2003, p. 50, based on a paper presented at the 2003 AIChE Spring National Meeting, March 30–April 3, 2003, New Orleans.

9. Huang, S., et al. "Select the Optimum Extraction Method for LNG Regasification," *Hydrocarbon Processing*, July 2004, p. 57.

10. Finn, C. I., G. L. Johnson, and T. R. Tomlinson. "Developments in Natural Gas Liquefaction," *Hydrocarbon Processing*, April 1999, p. 47, and C. Yost and R. DiNapoli. "Benchmarking Study Compares LNG Plant Costs," *Oil & Gas Journal*, April 14, 2003, p. 56.

11. Yost and DiNapoli. "Benchmarking Study Compares LNG Plant Costs," April 14, 2003, p. 56.

12. Flower, A. "Market Access Remains Key for LNG Producers," *Oil & Gas Journal*, April 22, 2002, p. 74, quoting A. Avidan (Bechtel Inc.), Gastech 2000, Houston.

13. Avidan, A., D. Messersmith, and B. Martinez. "LNG Liquefaction Technologies Move Toward Grater Efficiencies, Lower Emissions," *Oil & Gas Journal*, August 19, 2002, p. 60.

14. Kleiner, F., S. Rausch, and J. Knabe. "Increase Power and Efficiency of LNG Refrigeration Compressor Drivers," *Hydrocarbon Processing*, January 2003, p. 67.

15. Avidan, A., B. Martinez, and W. Varnell. "Study Evaluates Design Considerations of Larger, More Efficient Liquefaction Plants." *Oil & Gas Journal*, August 18, 2003, p. 50.

16. Woodside Petroleum Ltd. "1998 Annual Report." This project taps a condensate-rich field. Other projects such as Snohvit have lower LPG yields ("Construction of Most Northerly LNG Project Starts," *Oil & Gas Journal*, November 25, 2002, p. 38).

17. Masseron, J. "Petroleum Economics," *Editions Technip*, 1990, pp. 159–213, gives a good exposition of oil transport costs and LNG costs by tanker.

18. Capital cost is estimated from A. Kaplan, et al. "2003 LNG World Trade and Technology," wall chart, *Oil & Gas Journal*, PennWell, November 2003, quoting Cotton & Co. data.

19. Foglietta. *Hydrocarbon Processing*, January 2004, p. 39.

20. European Standard EN1473. "Installation and Equipment for Liquefied Natural Gas—Design of Onshore Installations," 1997.

21. Agrawal, D. "Select the Correct LNG Storage Tank for Your Facility," *Hydrocarbon Processing*, January 2004, p. 61.

22. McCall, M. M. "Salt Caverns Bring Security, Economy and Capacity to LNG Receiving," *Hydrocarbon Processing*, January 2004, p. 59.

23. Kaplan, A., et al. "2003 LNG World Trade and Technology," wall chart, *Oil & Gas Journal,* PennWell, November 2003.

24. Price, B. C. "Optimize Energy Integration for LNG Terminals," *Hydrocarbon Processing*, July 2004, p. 43.

25. Kaplan, A. "2003 LNG World Trade and Technology," quoting Foster Wheeler.

26. Neil, C. "Distance Continues to Drive LNG Costs for U.S. Delivery," *Oil & Gas Journal*, May 17, 2004, p. 56, quoting U.S. EIA, "The Global Liquefied Natural Gas Market, Status and Outlook," DOE/EIA–0637, December 2003.

27. Wagner, J. V., and S. van Wagensveld. "Marine Transportation of Compressed Natural Gas: A Viable Alternative to Pipeline or LNG," SPE 77925 presented at SPE Asia Pacific Oil & Gas Conference & Exhibition, Melbourne, October 8–10, 2002, abridged as "Marine Transportation of Compressed Natural Gas," in *Journal of Petroleum Technology*, September 2003, p. 73.

28. Wagner, J., and S. Cone. "Analysis Says Gas FPSO Feasible, CNG Possibly Economic Export Option," *Oil & Gas Journal*, December 2, 2002, p. 68.

12

Ammonia and Derivatives

This chapter will consider the use of natural gas for the production of ammonia and ammonia derivatives that contain nitrogen. The principal interest is the production of ammonia itself. Ammonia is increasingly used as a fertilizer, especially for the production of cereal crops, and there is a growing international trade in liquid ammonia. However, the manufacture of the nitrogen fertilizers urea and ammonium nitrate takes up most of the ammonia produced. Ammonium nitrate has an additional use as a blasting explosive and is used in large quantities by the minerals extraction industries.

In addition to these large-scale commodity chemicals, the production of hydrogen cyanide will also be considered. Hydrogen cyanide is used extensively for the extraction of gold and the production of nitrogen-containing chemicals such as acrylonitrile and polyamides.

Ammonia Manufacture

Figure 12–1 illustrates the principal unit operations for making ammonia from natural gas. The first step is steam reforming to convert most of the gas into synthesis gas. The second step (secondary reforming) is partial oxidation to complete the conversion of the methane in the gas. Air is used in this step, and this introduces nitrogen into the process stream. (These issues were discussed in chapter 7.)

Several steps aimed at maximizing the hydrogen (WGS) and removing all of the carbon present as carbon dioxide follow secondary reforming. Methanation removes the last traces of carbon oxides. The gas stream is then compressed to the high pressure (250 bar) required for ammonia synthesis. The ammonia synthesis reaction proceeds at low pass conversion, and there is a large recycle stream. The liquid ammonia is passed to storage in a pressure vessel.

Ammonia manufacture has been practiced for almost 100 years. Because of its importance in the manufacture of explosives and fertilizers, there are many plants (more than 1,000) and many designs. There also are many small plants producing less than 200 t/d of ammonia, and these are often used to supply a small, but captive, market.

Nowadays, most interest is in the production of large-scale plants for the international trade. Extensive improvements have resulted in production increases to more than 1,000 t/d ammonia for such plants.[1] The various technologies are differentiated by the methods through which the heat required by the steam reformer and the heat generated by the secondary reformer are integrated. They are also differentiated by the design of the converter to maximize the pass yield.[2]

Ammonia is stored and shipped in large pressure vessels. Some also use cryogenics. Ammonia shipping is tied into the LPG shipping trade, and many ships are constructed to carry either LPG or ammonia.

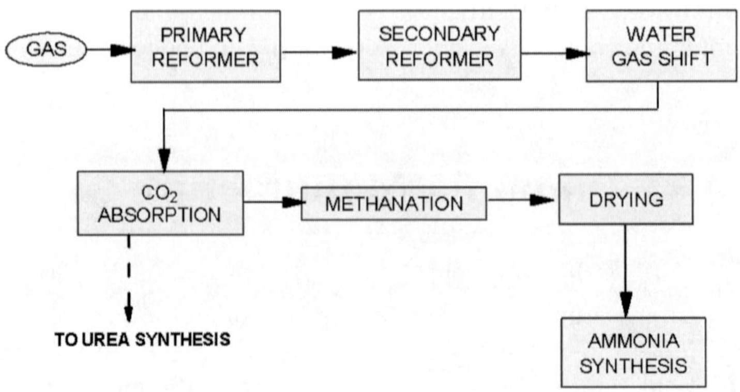

Fig. 12–1. Ammonia synthesis from natural gas

Ammonia Derivatives: Urea and Ammonium Nitrate

The major portion of ammonia manufacture is concerned with the production of derivatives, principally urea and ammonium nitrate. These are solid materials and easier to ship in large bulk carriers.

Urea

Urea is made by reacting the carbon dioxide produced in the manufacture of ammonia with ammonia at temperature and pressure (fig. 12–2).

$$2NH_3 + CO_2 = NH_2.CO.NH_2 + H_2O \qquad (12.1)$$

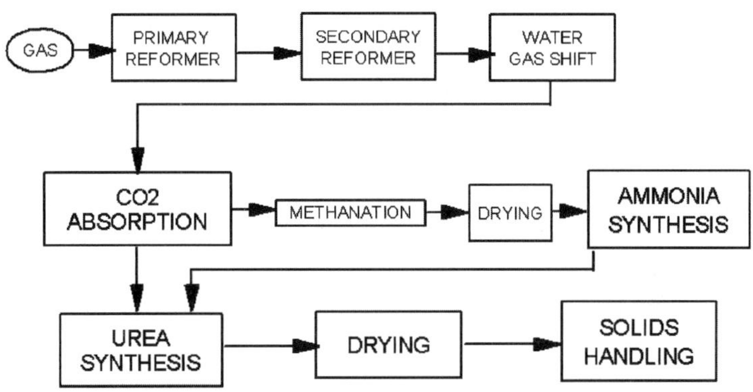

Fig. 12–2. Urea synthesis

The solid product is dried, and the powder is fabricated into granules or prills. These are sold in bags or as bulk solid for transport in ships. Urea has some uses in the chemicals industry (e.g., to produce urea formaldehyde resin). However, its principal use is as a fertilizer offering a nitrogen content of 46%.

Ammonium nitrate

Ammonium nitrate is produced in two stages by oxidation of ammonia to nitric acid.[3] This is subsequently reacted with more ammonia to give the following products (fig. 12–3):

$$NH_3 + 2O_2 = HNO_3 + H_2O \tag{12.2}$$

$$HNO_3 + NH_3 = NH_4.NO_3 \tag{12.3}$$

The solid products are prilled and then bagged or shipped in bulk. Ammonium nitrate has a use for both fertilizer and as a constituent of slurry-based blasting explosives. The latter is admixed with diesel fuel in the minerals extraction industry. Ammonium nitrate use is banned in many parts of the world because of its explosive nature.

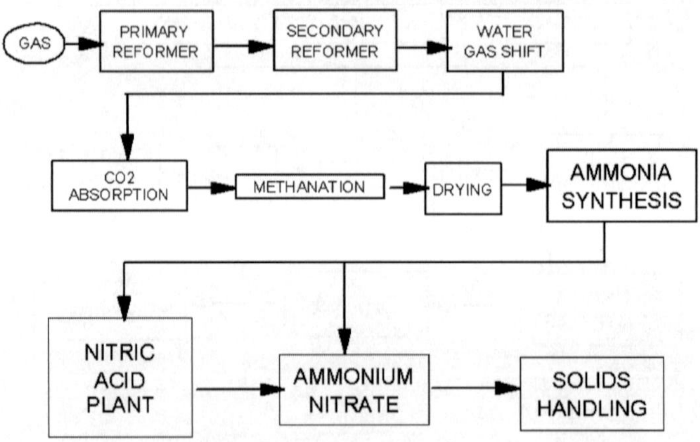

Fig. 12–3. Ammonium nitrate synthesis

Ammonium sulphate. Another important contributor to the nitrogenous fertilizer market is ammonium sulphate. This is made by simply adding ammonia (as an aqueous solution) to sulfuric acid. Large volumes are produced as by-products from scrubbing waste streams containing sulfuric acid with ammonia, or vice versa.

Complex fertilizers. A growing sector of the fertilizer trade is the production of complex fertilizers. These products combine a nitrogenous fertilizer (urea or ammonium nitrate) with a source of potassium (potash) and phosphorus. An attractive component is diammonium phosphate. This is made from ammonia and phosphoric acid, usually close to a source of phosphate rock.

Major Markets for Ammonia and Derivatives

Most fertilizer is used in the production of grasses and cereal crops.[4] Figure 12–4 illustrates the dominance of cereal crops in the use of fixed nitrogen fertilizer.

World production of ammonia and its derivatives was about 109 MMt in 2004. However, because of its importance to the agricultural industries, most production is for protected or regulated markets. Only about 20% is considered as free trade. But even here there can be nonmarket issues that favor one supplier over another.

Most fertilizer is shipped in bulk carriers as urea. However, there is an extensive and growing trade in ammonia as the anhydrous liquid. It is transported in pressure vessels by truck, barge, and ship in quantities up to about 75,000 t. (The carriers can often also transport LPG.) It is also transported by pipeline. Transport costs were discussed in chapter 5.

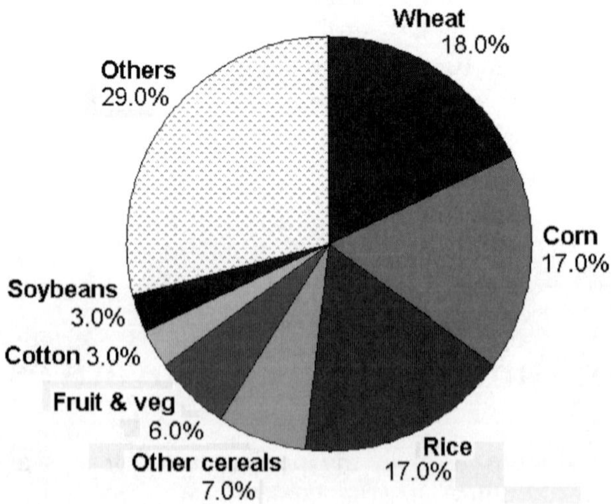

Fig. 12–4. World fertilizer consumption by crop

World nitrogen fertilizer trade

There has been a steady rise in the use of nitrogen fertilizers over the past decade. The global picture masks certain dynamic changes in the market. There has been a recent trend in declining nitrogen use in the OECD countries and in Eastern Europe. This has been more than offset by the rise in fertilizer use in the Far East. The outputs of the major producers are shown in figure 12–5.

The largest production is in China, where the principal feedstock is coal. China, like most other countries, uses most of its fertilizer on the domestic market. A few countries that have access to low-cost gas (Russia, Saudi Arabia, Trinidad, Canada, and Indonesia) dominate the international trade.

Historically, the major market player in the international trade has been Russia, with export through the Ukraine Black Sea ports. Russia possesses a large merchant marine dedicated to shipping ammonia, which requires cryogenic tankers.

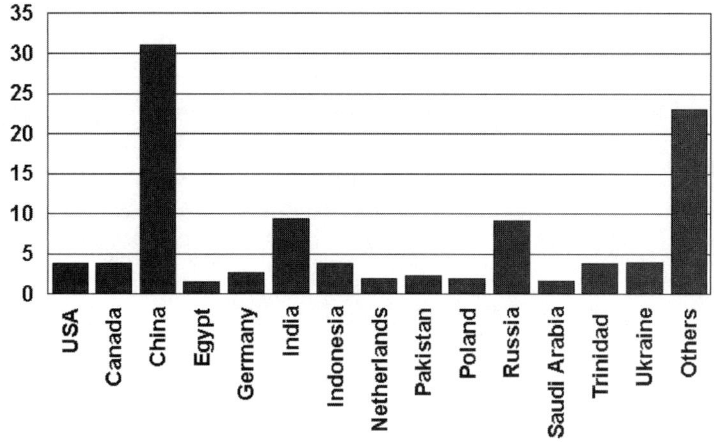

Fig. 12–5. 2004 world fixed nitrogen production (million tonnes). Data from D. A. Kramer, U.S. Geological Survey, Mineral Commodity Summaries, 1995.

Ammonia Prices and Price Differentials

The principal interest is the price of fixed nitrogen to the farmer. On a pure compound basis, the relative amounts of nitrogen in the main materials are: ammonia (82.5%), urea (46.6%), ammonium nitrate (35%), and ammonium sulphate (21.2%). Prices generally reflect these differences in the nitrogen levels. However, differentials between the various forms of nitrogen fertilizer are the rule. Market forces determine the difference. Generally ammonia is the cheapest form of nitrogen, and this has increased its market share in recent years.

The traded price of ammonia and urea, like all other traded goods, depends on the supply/demand balance. The supply/demand balance depends on geographic location and season. Because of its size, the movement of nitrogenous fertilizer in the United States dominates world trade and prices. There are major movements by barge along the Mississippi, and an extensive network of pipelines delivers liquid ammonia to the corn belt states. Imports are from two main sources: by truck and rail from Canada and by ship into seaports along the Gulf Coast.

At any one time there can be marked differentials in the price of ammonia and urea depending on location. Lowest prices occur in the regions of production and are highest for the areas of import. For example, there is a sharp contrast in price between the U.S. Gulf Coast (which has both manufacturing plants and large import terminals) and California. The European price is broadly in line with those for the U.S. Gulf price, having both producers and large import terminals. The lowest prices (fob) are usually at Black Sea ports and at major producers, such as Trinidad in the Caribbean. Trinidad produces large quantities of ammonia for the U.S. market.

U.S. fertilizer: seasonal price changes

Because the U.S. market is so large, the seasonal demand for fertilizer results in price fluctuations elsewhere in the world. Figure 12–6 illustrates the seasonal nature of prices for ammonia on the U.S. market for the period in the early 1990s. (This was a time of relatively stable prices, and the seasonal influence of the U.S. market can be clearly seen.)

The seasonal demand changes in the United States impact the availability of the merchant marine. Because the ammonia fleet also services the ocean transport of LPG, the price of LPG carriage varies seasonally.

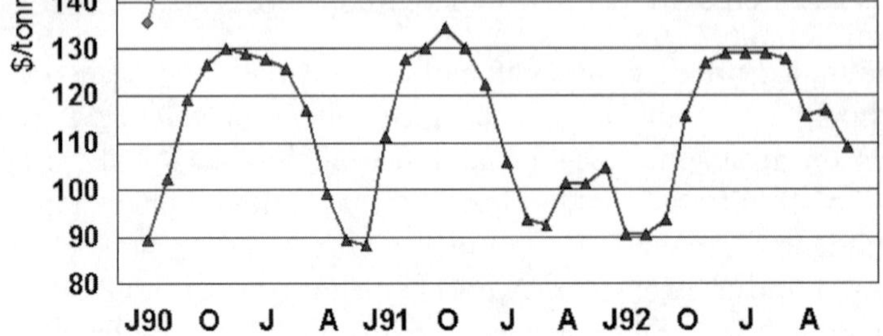

Fig. 12–6. U.S. seasonal price variations (July 1990 to July 1992)

Ammonia price: historical trends

Figure 12–7 illustrates the variation in landed ammonia price [cost and freight (C&F) Europe] since 1988.

The figure can be considered in three periods. The early period (until about 1994) was characterized by low and relatively stable prices. Prices ranged from about \$100/t to about \$160/t and averaged about \$130/t. Variation can be explained in terms of the seasonal demand in the U.S. market. This was a time when the trade was dominated by the USSR's (Russia's) supply. The USSR's command economy controlled production in Russia, transport to the export terminals on the Black Sea (Ukraine), and shipping in the Soviet merchant fleet.

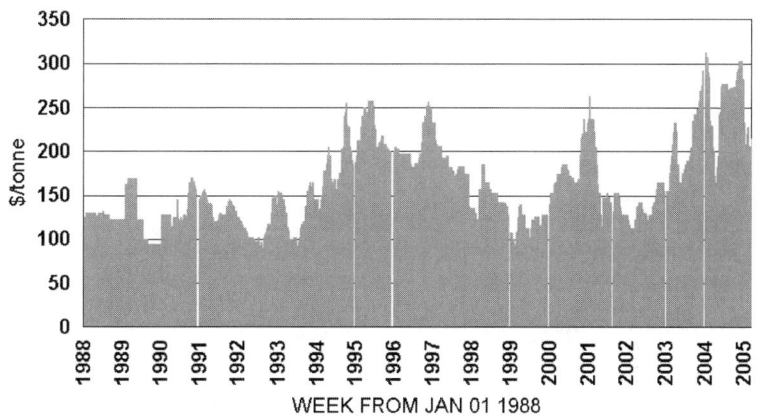

Fig. 12–7. Historical ammonia prices

From about 1994 to about 2002, there was a rapid rise and an increase in the price (average about \$150/t) and volatility (range \$100/t to more than \$250/t). This can be explained by production difficulties as a consequence of the breakup of the USSR and the rapidly rising demand for anhydrous ammonia exacerbating market oversupply and shortfalls.

Since 2002 there has been a steady rise in the price of oil. As a consequence, the price of natural gas has risen in many of the producing and exporting countries. This has resulted in a steady rise in the price of ammonia, with prices peaking at more than \$300/t in early 2004 and 2005.

Ammonia and Urea Production Costs

The cost of ammonia production is strongly dependent on the capital cost and the cost of feed gas. The feed gas usage has been estimated from the claimed efficiencies of modern plants to be about 7 Gcal/t or 25 MMBTU/t on an LHV basis. It has been converted to HHV through the addition of an operating allowance of 3% plus an allowance for catalyst deterioration of 5%. The gas usage for urea is considered the same as ammonia, since there is often an excess of energy available from the ammonia synthesis step.

The fixed variable relationship for the production of ammonia and urea is shown in figure 12–8, with the relevant statistics given in table 12–1.

A comparison of figures 12–8 and 12–7 illustrates that for a viable plant, gas supply has to be below about $1.50/GJ. Further, most export plants are located in countries with considerably higher construction costs than those of the U.S. Gulf Coast region. Thus the gas prices for a viable plant would be less than $1/GJ.

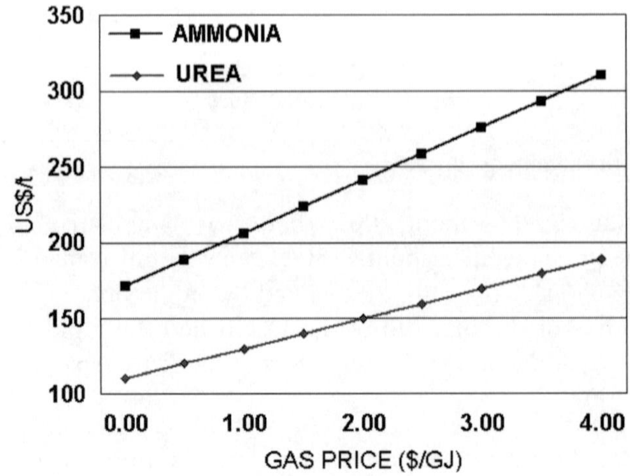

Fig. 12–8. Ammonia and urea production costs

Table 12–1. Statistics for ammonia and urea production

		Ammonia	Urea
Production	kt/a	850	1500
Capital Cost	$MM	605.71	681.42
Construction period	years	3.00	3.00
Plant life	years	15.00	15.00
Return on capital	%/a	16.34%	16.34%
Non gas operating costs	$MM/a	145.95	165.53
Gas usage	PJ/a	29.62	29.62

Cyanide

The use of hydrogen cyanide is central to the production of gold and is used in the manufacture of certain types of nylon, methacrylic esters (Perspex), and other chemicals. Hydrogen cyanide can be produced from several sources, including natural gas by reaction with ammonia. There are several technologies. The Andrussov process is typical of processes using ammoxidation of methane, employing a platinum-rhodium gauze catalyst at 1,000°C–1,200°C.

$$CH_4 + NH_3 + 3/2O_2 = HCN + 3H_2O \qquad (12.4)$$

The reaction is exothermic (ΔHr = –473 kJ/mol). The product is rapidly quenched, and the hydrogen cyanide is obtained by distillation of the aqueous solution. There is some oxidation of methane to carbon dioxide. However, the selectivity is claimed to be 88% on methane and 90% on ammonia.

The Degussa BMA process is performed in the absence of oxygen and a product free of carbon dioxide is obtained.

$$CH_4 + NH_3 = HCN + 3H_2 \qquad (12.5)$$

This reaction is endothermic (ΔHr = +251 kJ/mol) and so requires a source of additional heat. The process uses sintered corundum tubes impregnated with platinum, rhodium, or aluminium at 1,200°C to 1,300°C. The process is similar to the operation of a steam reformer.

The ammonia reacts to completion, with any excess being removed by treatment with sulfuric acid. Product selectivity of 90% to 91% on methane and 83% to 84% on ammonia are claimed.

However, published data on the requirements for hydrogen cyanide production facilities indicate much lower selectivity than the above values.

Production economics

Data have been estimated from the Australian Gold Resources facility at Kwinana. This plant produces sodium cyanide by neutralizing the hydrogen cyanide produced with sodium hydroxide (caustic soda). The capital is estimated from the published cost for doubling the plant's capacity in 1998. This is taken as equivalent to one-third of the cost of a greenfield plant. The plant has the statistics given in table 12–2.

Table 12–2. Statistics for sodium cyanide production

Capacity (tonne dry NaCN)	65,000
Capital ($MM, US Gulf)	96.5
Construction period (y)	2
Non Feed Opex ($MM)	39.2
Gas Usage (GJ/t)	36
Sodium Hydroxide (t/t)	0.95
Ammonia (t/t)	0.65
Power (MW/t)	0.67

The cost of production depends not only on the cost of gas, but also on the cost of ammonia and sodium hydroxide. The latter two are usually imported to the plant, although some integrated chemicals facilities may have on-site ammonia and sodium hydroxide production. Using a value of $300/t for sodium hydroxide and $250/t for imported ammonia, the sensitivity to gas price is illustrated in figure 12–9.

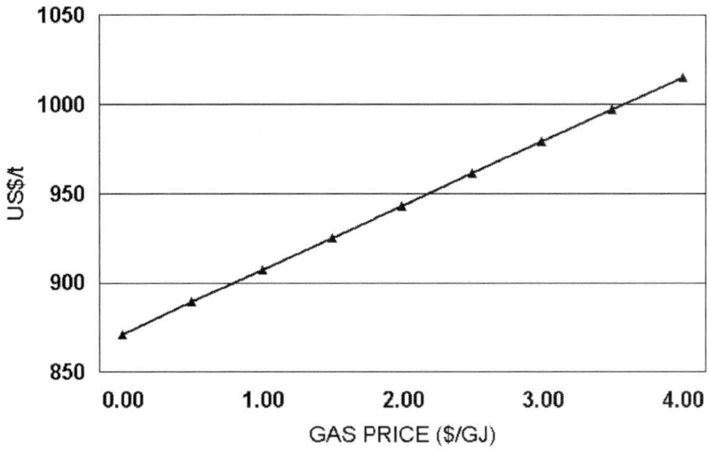

Fig. 12–9. Production cost of sodium cyanide

Sodium cyanide is available at prices in the region of $1,000/t. The figure indicates that the production cost is lower than this value at gas prices below $4/GJ.

References

1. "Petrochemical Processes 2003," *Hydrocarbon Processing*, March 2003, pp. 74–77, gives details of six proprietary technologies.

2. Jennings, J. R., and S. A. Ward. "Ammonia Synthesis," *Catalyst Handbook,* second edition, M. V. Twigg, editor, Wolfe Publishing, 1989.

3. Davies, P., R. T. Donald, and N. H. Harbord. "Catalytic Oxidations," *Catalyst Handbook,* second edition, M. V. Twigg, editor, Wolfe Publishing, 1989.

4. Data from Potash Corporation Web site: www.potashcorp.com

13

Ore Processing

Natural gas is used widely in the processing of ores, often acting as an energy source for the generation of electricity. This use was discussed in earlier chapters. This chapter will discuss the use of gas as a chemical reducing agent for the conversion of ores into metals, using the production of DRI as an example. It will also address the use of gas in alumina refining to provide heat and power for the Bayer process.

Direct Reduction of Iron Ore

Directly reduced iron (DRI), also known as sponge iron or hot briquetted iron (HBI), is a widely manufactured and traded commodity. This process of manufacture is more than 100 years old. It involves the reduction of iron ore with a reducing agent (coal or natural gas) without the formation of molten metal, hence its name. Coal is used directly as the reducing agent, whereas natural gas is first converted into a synthesis gas, often called *reducing gas*. DRI competes with pig iron, produced from blast furnace operations, and scrap iron. Its main use is as a feed for electrolytic steel making and mini-mill operations.

There are more than 100 variants in the technology.[1] Processes are often designed to suit a specific site need. Some generic types are listed in table 13–1. Most are continuous processes, although some operate in the batch mode. One pertinent issue is the choice of ore; some processes require lump ore or iron ore pellets, while others operate with iron ore fines.

Table 13–1. Some technologies for direct reduction of iron ore

Coal Based (lump ore or pellets)	Gas Based—Shaft (lump ore or pellets)	Gas Based—Fluid Bed (ore fines)
Continuous Processes		
SL/RN (Lurgi)	MIDREX	HIB
CODIR (Krupp)	WIDBERG	FIOR
SDR (Sumitomo)	ARMCO	FINMET
SPM (Sumitomo)	NIPPON STEEL	
KAWASAKI	ACCAR	
DRC		
Batch Process		
HOGANAS	HYL	

The competitive advantage for DRI is that it is economic at levels of production of 1 MMt/y and below. This is well below the economic viability for the operation of a modern blast furnace. Thus it can be sized to supply a local need. Most DRI plants operate at production rates of 350 t/y to 500,000 t/y DRI, but some have production levels greater than 2 MMt/y. Another advantage of increasing importance is that DRI avoids the use of coke ovens, which are nowadays considered to cause excessive environmental pollution.

The world production of DRI is about 55 MMt/y (4% of crude steel production). Interestingly, the technology is not widely used in the older economies of the United States, Canada, and Europe, where blast furnaces still dominate iron production. These economies account for only about 2 MMt/y (fig. 13–1).

Although the number of operations is split evenly between coal and natural gas, natural gas based plants are generally larger and account for most of the world's production.

About 64% of world production uses the Midrex process. Midrex reports that coal-based processes account for about 12% of the world production.[2]

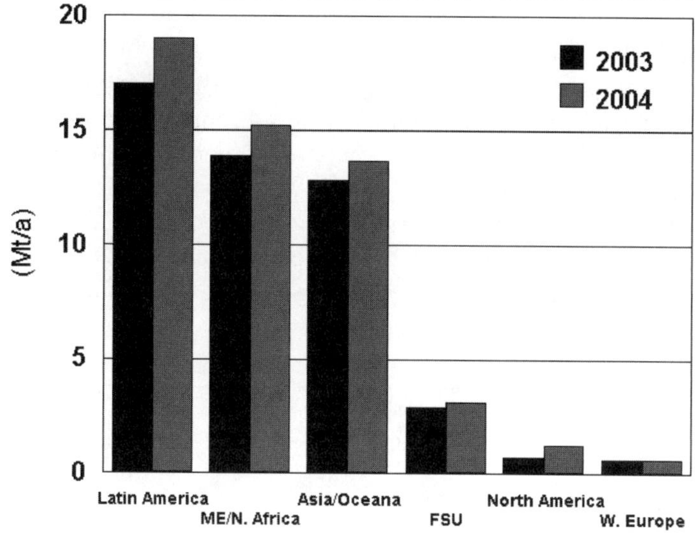

Fig. 13–1. Regional production of DRI

Overview of technology

The principal unit operations for the reduction of iron ore by natural gas are shown in figure 13–2. There are three main parts to the plant. Gas is reformed into a reducing gas, which is used to reduce the iron ore in the furnace. If the product after reduction is an aggregate of about 1 cm or larger, it is generally known as DRI. If the aggregate is smaller, it is passed to a briquetting plant, which produces HBI as the final product.

Natural gas is desulfurized and converted into synthesis gas, generally referred to as reducing gas, by steam reforming. The process is very similar to that for the production of synthesis gas for methanol manufacture. However, the amount of methane and carbon dioxide in the synthesis gas is minimized by steam reforming at higher temperatures and lower pressures than for methanol synthesis gas. This is to allow optimum performance in the reduction step. The reducing gas is fed to the iron ore reduction furnace. The principal reducing reactions are:

$$Fe_2O_3 + 3H_2 = 2Fe + 3H_2O \tag{13.1}$$

and

$$Fe_2O_3 + 3CO = 2Fe + 3CO_2 \tag{13.2}$$

These reactions are reversible at the operation temperatures. The concentration of hydrogen in particular has to remain high, hence low gas utilization. Also, the water content and the carbon dioxide content of the reducing gas must be maintained as low as possible. Water is easily removed by cooling the reducing gas stream, and in some processes, carbon dioxide is stripped from the gas. The low gas utilization makes it important to effectively integrate the reducing furnace and the reformer. Only about 30% to 40% of the reducing gas is used due to thermodynamic constraints in the reduction. The unused reducing gas is recycled to the reformer as feedstock or fuel.

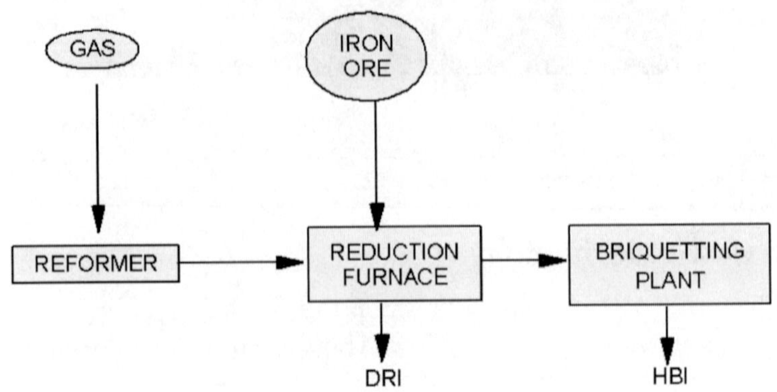

Fig. 13–2. Principal unit operations of DRI

For some processes, the reduction is carried out in a series of furnaces, the ore slowly passing from one furnace to the next until the desired level of reduction is achieved.

The product from the reduction furnace (> 93% metallized), is often pyrophoric. It is passivated by briquetting for transport and export. There are two generic technologies: shaft furnace and fluid-bed process.

Shaft-furnace type

The most common processes considered are those of Midrex and Lurgi. Both involve shaft-furnace reduction of lump or pelletized ore. The layout is illustrated in figure 13–3.

The iron ore is preheated before entering the top of the shaft furnace via a hopper system. As the ore descends the shaft, it is reduced by the rising reducing gas provided by the reformer. DRI leaves the bottom of the shaft. As gases leave the top of the shaft, they are cooled to remove water produced in the reduction process before recycling or using for reformer fuel.

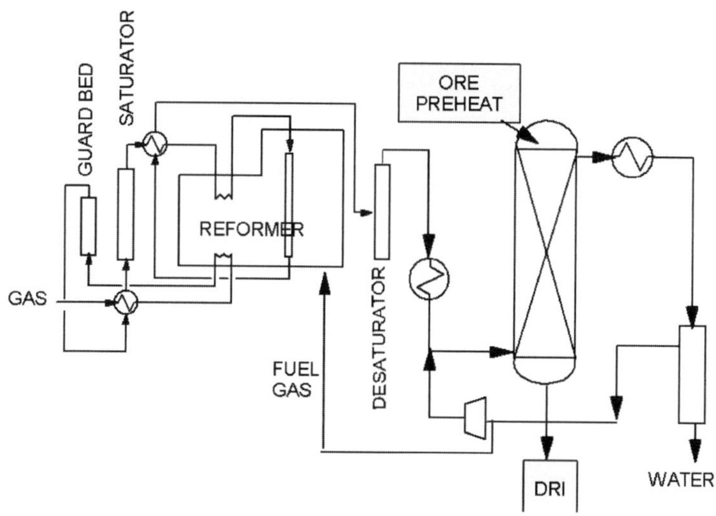

Fig. 13–3. Shaft-furnace unit operations

Fluid-bed process

Fluid-bed processes (e.g., FINMET and FIOR) use iron ore fines directly. The FINMET process was developed by FIOR de Venezuela and Voest-Alpine Industries of Austria. It uses ore fines and has been used by BHP-Billiton for its large HBI operation in Australia. [The BHP-HBI venture suffered a series of problems, including cost blowouts and operational problems, which lowered the output. At the time of writing (May 2005), the project was on a care-and-maintenance basis, with the carrying value fully written off in the BHP-Billiton accounts.]

The unit operations for the FINMET process are illustrated in figure 13–4.

Gas is reformed into reducing gas. This is then mixed with recycle gas and passed to the carbon dioxide and water removal units. The gas is then heated to the required temperature and passed to the lower of four reducing reactors. Off gases from reactors are passed to the next reactor in the sequence. The ore is dried in a fluid-bed drier using waste heat from the reformer. The ore then passes to the top of the first reactor and slowly progresses down the reactors in a countercurrent manner to the reducing gas flow. After the final reactor, the reduced product is passed to a briquetting plant.

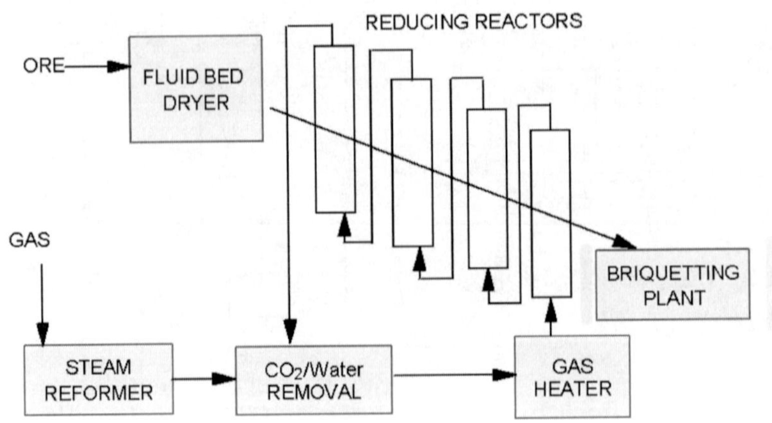

Fig. 13–4. FINMET process—unit operations

Market for DRI

DRI competes with high-quality ferrous scrap and pig iron, which are used to produce steel in electric arc furnaces (EAF). This method of steel manufacture can operate successfully on a small scale in mini-mills that usually produce a single product (e.g., concrete reinforcement bars) for a local market.

Steel can be considered as a mature product, and therefore its usage can be considered as proportional to a nation's GDP. The growth in steel usage will be proportional to the growth in GDP. For developing nations, where the GDP growth is very high by world standards, one can expect the increasing demand for steel to be even higher. This is because of the intensity in construction and infrastructure development, and hence an attraction for DRI/EAF projects.

In addition, DRI is being used in increasing quantities by conventional blast-furnace steel operations in order to increase capacity. Blast furnaces operate at a relatively large fixed output (e.g., 1 MMt), and small increases in productivity (100,000 t) are difficult. Adding DRI to the blast furnace feed can achieve this objective, thus avoiding the capital cost of upgrading and commissioning a new blast furnace.

DRI use in blast furnaces also has positive environmental effects that are gaining it increased interest, especially by steel makers in the United States. This is because the DRI needs very little coke to convert it into pig iron. Consequently, the use of DRI (without an overall capacity increase) results in less demand on the most polluting section of a steelworks, the coke ovens. In environmentally conscious nations, coke oven emissions are severely regulated. The use of DRI in the blast furnace offers steel makers the opportunity to achieve legislated standards with relatively old plant formats. The use of DRI in blast furnace operations potentially offers the sale of very large volumes of DRI.

The production of DRI has increased dramatically over the past two decades. The growth is illustrated in figure 13–5.

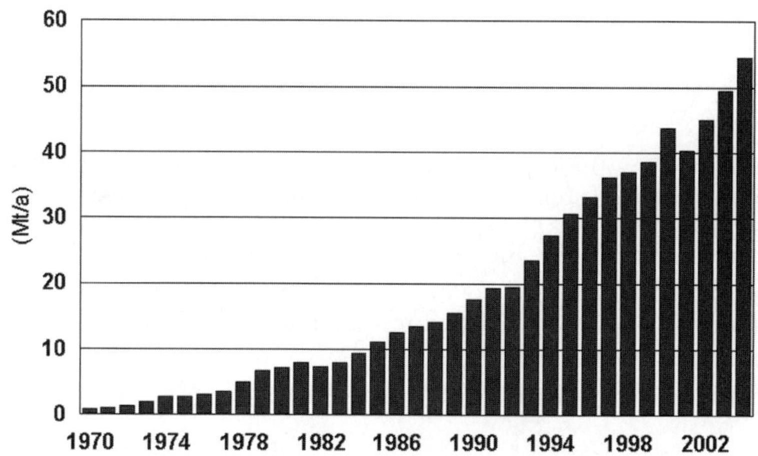

Fig. 13–5. Historical growth in DRI production

Prices

The value of DRI is related to that of the alternatives, particularly merchant premium scrap iron and steel. Because DRI contains few *tramp* metals, it usually sells at a premium of about \$30/t over scrap. The price of scrap varies with region and according to the business cycles. Figure 13–6 shows historical scrap prices for the United States.

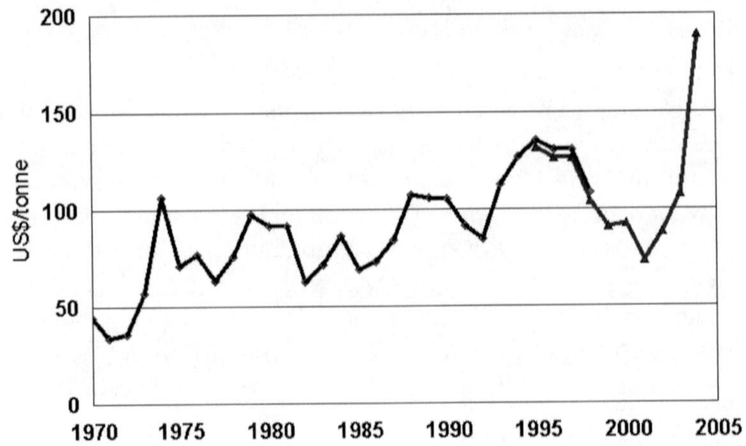

Fig. 13–6. Historical U.S. scrap prices. Source: M. Fenton, U.S. Geological Survey, Mineral Commodity Summaries. Note that two series are used, hence the discontinuity in the graph.

For developed nations, where there is usually a large volume of scrap available, prices are typically \$100/t. In developing countries where scrap has to be imported, the price can be considerably higher, from \$150/t to \$180/t, making DRI at about \$200/t competitive.

In recent times, due to the unprecedented demand for steel products in China, scrap prices have increased dramatically over these historical prices. The U.S. scrap price approached \$200/t in 2004.

Costs of production

Obviously, production costs depend on the ore and gas prices. The costs of iron ore are usually settled on an annual basis between major suppliers in Australia, Brazil, and Africa, with the major steel makers in Japan and Europe. These annual price negotiations set the price for the market. About 60% of the market is for ore fines. Lump and pellet prices are then determined as a premium to the fines ore price. A percentage change to the previous year's pricing structure is negotiated. Then iron ore prices are denominated in U.S. cents per dry metric tonne unit, reflecting a common base unit for iron content. The ore price is obtained by multiplying this value by the iron content of the ore.[3] Current prices for ore are given in table 13–2.

Table 13–2. Current iron ore prices

	2004	2005
U.S. cents/dry metric tonne		
Lump ore	45.93	78.77
Fine ore	35.99	61.72
Ore prices @60% Fe US$/t		
Lump ore	27.558	47.262
Fine ore	21.594	37.032

The production cost statistics shown in table 13–3 have been developed from the BHP HBI plant using the FINMET process at Port Hedland in Australia. The values taken from BHP-Billiton published accounts have been adjusted to reflect a U.S. Gulf Coast operation. A second smaller option is evaluated to represent a smaller operation using lump ore as a feedstock.

Table 13–3. Statistics for DRI production

		Large	Small
Production	kt/a	2000	200
Capital cost	$MM	904.7	180.5
Construction period	years	3	3
Plant life	years	15	15
Return on capital	%/a	16.34%	16.34%
Non gas operating costs	$MM/a	312.16	55.62
Gas Usage	PJ/a	44.2	4.42
Ore Cost	$/t	37	47.3

The variable relationship of DRI against gas price is illustrated in figure 13–7.

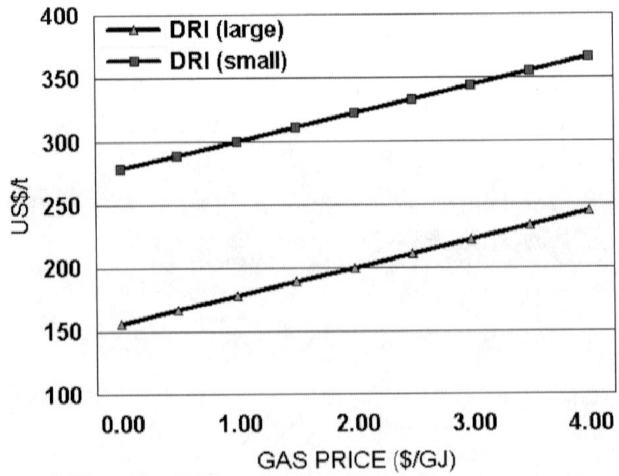

Fig. 13–7. DRI unit production costs

The large-scale operation based on the BHP-HBI system produces 2 MMt of DRI from about 2.5 MMt of iron ore fines. These are priced at $37/t to reflect the 2005 price. The scaled-down version of 200,000 t/y gives an estimate for the costs in developing countries using imported lump ore at $47/t.

The fixed-variable relationship shows that DRI would be competitive with scrap if gas prices were less than about $2/GJ for large-scale export projects. For the smaller plant, the plant profitability will depend on the local price of steel as well as the availability of low-cost gas.

Alumina

Alumina is produced from bauxite by the Bayer process. Large scale alumina refineries demand large amounts of energy for power generation and for evaporation of the alumina slurries. Alumina refineries often involve large-scale cogeneration schemes (see chapter 6).

Figure 13–8 illustrates the principal unit operations involved in the Bayer process for converting bauxite into alumina. Bauxite, lime, and caustic soda enter the cycle in the digester, which is a series of autoclave or tube reactors held between 120°C and 250°C. The quantities of caustic soda and lime and the details of the operation are dependent on the nature of the bauxite, which varies according to the ore body. The digester produces a solution of sodium aluminate and a suspension of undissolved high iron content solids. This is cooled, and the solids, known as *red mud*, are separated in the clarifier. Red mud is further washed to remove excess caustic before being disposed. About 1–2 t of red mud are produced per tonne of alumina.

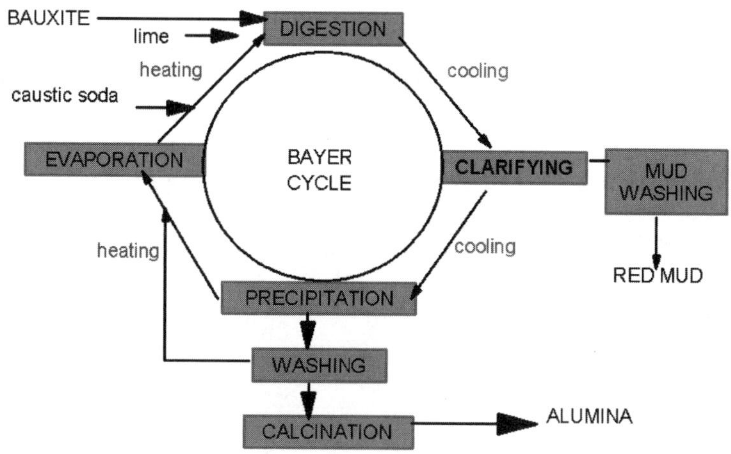

Fig. 13–8. Alumina manufacture by the Bayer process

Upon further cooling, alumina trihydrate separates. This is washed, and the washing solution is returned to the cycle. The alumina trihydrate is then calcined to produce alumina. The alkaline solution is concentrated by heating in the evaporator before returning to the bauxite digester.

Energy is required in the cycle to heat and evaporate solutions and for mechanical agitation of the reacting slurry. More energy is required in the calcination unit to produce alumina from the primary product (aluminium trihydrate) of the Bayer process.

Alumina refineries are very large operations, each year producing 1 MMt or more of alumina. This requires 30 to 50 PJ of energy per annum. The requirement for heat and electricity for pumping and stirring makes alumina refineries ideal candidates for large-scale cogeneration projects.

Economics

An alumina refinery producing 1.4 MMt of smelter-grade alumina was opened in Gladstone, Australia in 2005.[4] This refinery cost AUS$1.5 billion (US$1.125 billion at AUS$/US$ = 0.75). Although this refinery is driven by a coal-fired power station, a similar cost would be expected for a gas-fired refinery.

References

1. Stephenson, R. L., and R. M. Smaller, editors. *Direct Reduced Iron—Technology and Economics of Production and Use*, Warrendale, PA: Iron and Steel Society, 1980.
2. Midrex Web site at www.midrex.com
3. Rio Tinto Limited. "Iron Ore Price Settlement," News release, February 23, 2005.
4. Comalco. "Comalco Alumina Refinery Officially Opens," Media Release, March 4, 2005.

Appendix A

Abbreviations

ALPO	alumina and phosphate molecular sieve
atm	atmosphere (pressure)
AVTUR	aviation turbine fuel; jet fuel
bbl	petroleum barrel
Bcf/d	billion cubic feet per day
bcm	billion cubic meters
BFW	boiler feed water
BTU	British thermal unit
°C	degrees centigrade (Celsius)
C&F	cost and freight
CBM	coalbed methane
cf	cubic foot
cif	container, insurance, and freight (destination port price)
CN	carbon number
CNG	compressed natural gas
DCF	discounted cash flow
DME	dimethyl ether
DRI	directly reduced iron
DWT	deadweight tonnes
EAF	electric arc furnaces
EDC	ethylene dichloride
EOR	enhanced oil recovery
EU	European Union
°F	degrees Fahrenheit
FCC	fluid cat cracker
FLNG	floating production of liquefied natural gas
fob	free on board (embarkation port price)
FPSO	floating production storage and off-take
FT	Fischer-Tropsch
FTR	Fischer-Tropsch reactor
GAIL	Gas Authority of India, Ltd.
GHG	greenhouse gas
GHR	gas-heated reforming
GJ	gigajoule
GTL	gas to liquids
HBI	hot briquetted iron
HHV	higher heating value (gross)
HP	horsepower
HSLA	high-strength low-alloy
HT shift	high-temperature shift
ISBL	inside battery limits
JV	joint venture
kW	kilowatt

kWh	kilowatt hour
L	liter
LAB	linear alkylbenzenes
lb	pound
LF	load factor
LHV	lower heating value (net)
LNG	liquefied natural gas
LPG	liquefied petroleum gas (usually propane and butane)
LT shift	low-temperature shift
m^3	cubic meter
Mcf	thousand cubic feet
MDS	middle distillate synthesis
MMcf	million cubic feet
MMBTU	million (U.S. customary) BTU
MMscfd	million standard cubic feet per day
MMt	million tonnes
$MMtCO_2e$	million tonnes of carbon dioxide equivalent
MOGD	methanol to gasoline and distillate
MRP	mixed refrigerant process
MTBE	methyl tertiary-butyl ether
MTFB	multitubular fixed bed
MTG	methanol to gasoline
MTO	methanol to olefins
MTP	methanol to propylene
NGE	natural gas expansion
NGH	natural gas hydrates
PJ	petajoule (1×10^{15} joules)
PSA	pressure swing absorption
psia	pounds of force per square inch absolute
PVC	polyvinyl chloride
ROC	return on capital
RON	research octane number
ROW	right of way
SAPO	silica, alumina, and phosphate molecular sieve
scf	standard cubic feet
SNG	synthetic natural gas (methane)
SPARG	sulfur-passivated steam reforming (SPARG)
SR	stoichiometric ratio
t	tonne or metric ton
Tcf	trillion cubic feet
Tcfd	trillion cubic feet per day
TEG	triethylene glycol
THT	tetrahydrothiophene
VCM	vinyl chloride monomer
WHB	waste heat boiler
WTI	West Texas Intermediate; a marker (price setting) crude oil
y	year

Appendix B

Useful Conversion Factors for Fuels and Products

Table B–1. Basic conversion factors

m^3	35.315	cf
m^3 @15°C	35.383	cf @60°F
GJ	0.9478	MMBTU
$/GJ	1.055	$/MMBTU
1 kWh	3.6	MJ
Lb	0.4536	kg
HP	0.7457	kW

Table B–2. Temperature conversions

	°C	°F
Absolute zero	–273.15	–459.67
Normal	15	59
STP (metric)	0	32
Standard	15.56	60

Table B–3. Specific volumes and heating values of liquid fuels

	L/t	bbl/t	HHV GJ/t
Ethane	2,654		
Propane	1,998		
Butanes	1,928	12.13	49.6
Naphtha	1,534	9.00	48.1
Gasoline	1,360	8.56	46.4
AVTUR	1,261	7.93	46.4
Diesel	1,182	7.43	45.6
Fuel oil (LS)	1,110	6.98	44.1
Crude oil (35°API)	1,177	7.40	45
Crude oil (40°API)	1,212	7.62	

Table B–4. Energy values of some products and intermediates

	HHV (GJ/t)	LHV (GJ/t)
Carbon Monoxide	10.1	10.1
Butenes	48.1	45
Ethylene	50.3	47.2
Hydrogen	141.8	120
Methanol	22.7	19.5
Propylene	48.9	45.8
DME	31.0	28.4
Carbon	32.8	32.8
Ammonia	22.5	18.6

Table B–5. Properties of some coals

Type		Wyoming	Witbank	Illinois No 6	Wyodak	Ger Brown
Ultimate Analysis (DAF)						
Carbon	wt%	74.45%	81.25%	78.10%	75.60%	67.50%
Hydrogen	wt%	5.10%	5.00%	5.50%	6.00%	5.00%
Oxygen	wt%	19.25%	10.00%	10.90%	16.80%	26.50%
Nitrogen	wt%	0.75%	2.50%	1.20%	0.70%	0.50%
Sulfur	wt%	0.45%	1.25%	4.30%	0.90%	0.50%
		100.00%	100.00%	100.00%	100.00%	100.00%
Ash (as received)	wt%			12.0%	5.9%	6.4%
Moisture (as received)	wt%			6.5%	35.0%	5.0%
As Received Basis						
Carbon	wt%			65.91%	53.66%	60.59%
Hydrogen	wt%			4.64%	4.26%	4.49%
Oxygen	wt%			9.20%	11.92%	23.79%
Nitrogen	wt%			1.01%	0.50%	0.45%
Sulfur	wt%			3.63%	0.64%	0.45%
Ash (as received)	wt%			10.13%	4.19%	5.75%
Moisture (as received)	wt%			5.49%	24.84%	4.49%
				100.00%	100.00%	100.00%
LHV (as received)	GJ/t			25.80	17.16	9.90
HHV (as received)	GJ/t			26.82	18.10	10.89
LHV (DAF)	GJ/t			30.57	24.18	11.03
HHV (DAF)	GJ/t	29.6	31.4	31.79	25.50	12.13

DAF is dry ash free basis of analysis

Appendix C

Cost of Utilities

Table C–1. Utility costs

Days per year		340
Hours per year		8,160
Electricity		
Purchases	¢/kWh	5.0
Export	¢/kWh	3.0
Steam		
High pressure	$/t	2.04
Medium pressure	$/t	1.81
Low pressure	$/t	1.36

Table C–2. Impact of oil price on fuel prices (Singapore)—US$/bbl

Oil (Tapis)	30.0	35.0	40.0	45.0	50.0	55.0
Naphtha	29.8	34.8	39.9	44.9	50.0	55.0
Gasoline	36.1	41.8	47.4	53.1	58.7	64.4
AVTUR	38.8	45.5	52.3	59.0	65.8	72.5
Diesel	35.6	41.4	47.1	52.9	58.6	64.4

Appendix D

Nelson-Farrar Refiner Cost Indices

Table D–1. Nelson Farrar refiner cost indices

Year	Material	Equip..	Labor	Index	NF Factor
1946	100.0	100.0	100.0	100.0	18.561
1947	122.4	114.2	113.5	117.1	15.856
1948	139.5	122.1	128.0	132.6	13.998
1949	143.6	121.6	137.1	139.7	13.286
1950	149.5	126.2	144.0	146.2	12.696
1951	164.0	145.0	152.5	157.1	11.815
1952	164.3	153.1	163.1	163.6	11.347
1953	172.4	158.8	174.2	173.5	10.699
1954	174.6	160.7	183.3	179.8	10.322
1955	176.1	161.5	189.6	184.2	10.077
1956	190.4	180.5	198.2	195.1	9.515
1957	201.9	192.1	208.6	205.9	9.014
1958	204.1	192.4	220.4	213.9	8.678
1959	207.8	196.1	231.6	222.1	8.358
1960	207.6	200.0	241.9	228.2	8.134
1961	207.7	199.5	249.4	232.7	7.976
1962	205.9	198.8	258.8	237.6	7.811
1963	206.3	201.4	268.4	243.6	7.621
1964	209.6	206.8	280.5	252.1	7.361
1965	212.0	211.6	294.4	261.4	7.100
1966	216.2	220.9	310.9	273.0	6.798
1967	219.7	226.1	331.3	286.7	6.475
1968	224.1	228.8	357.4	304.1	6.104

1969	234.9	239.3	391.8	329.0	5.641
1970	250.5	254.3	441.1	364.9	5.087
1971	265.2	268.7	499.9	406.0	4.571
1972	277.8	278.0	545.6	438.5	4.233
1973	292.3	291.4	585.2	468.0	3.966
1974	373.3	361.8	623.6	523.5	3.546
1975	421.0	415.9	678.5	575.5	3.225
1976	445.2	423.8	729.4	615.7	3.015
1977	471.3	438.2	774.1	653.0	2.843
1978	516.7	474.1	824.1	701.1	2.647
1979	573.1	515.4	879.0	756.6	2.453
1980	629.2	578.1	951.9	822.8	2.256
1981	693.2	647.9	1,044.2	903.8	2.054
1982	707.6	662.8	1,154.2	975.6	1.903
1983	712.4	656.8	1,234.8	1,025.8	1.809
1984	735.3	665.6	1,278.1	1,061.0	1.749
1985	739.6	673.4	1,297.6	1,074.4	1.728
1986	730.0	684.4	1,330.0	1,090.0	1.703
1987	748.9	703.1	1,370.0	1,121.6	1.655
1988	802.8	732.5	1,405.6	1,164.5	1.594
1989	829.2	769.9	1,440.4	1,195.9	1.552
1990	832.8	797.5	1,487.7	1,225.7	1.514
1991	832.3	827.5	1,533.3	1,252.9	1.481
1992	824.6	837.6	1,579.2	1,277.4	1.453
1993	846.5	842.8	1,620.2	1,310.7	1.416
1994	877.2	851.1	1,664.7	1,349.7	1.375
1995	918.0	879.5	1,708.1	1,392.1	1.333
1996	917.1	903.5	1,753.5	1,419.0	1.308
1997	923.9	910.5	1,799.5	1,449.2	1.281
1998	917.5	933.2	1,851.0	1,477.6	1.256
1999	883.5	920.3	1,906.3	1,497.2	1.240
2000	896.1	917.8	1,973.7	1,542.7	1.203
2001	877.7	939.3	2,047.7	1,579.7	1.175
2002	899.7	951.3	2,137.2	1,642.2	1.130
2003	933.8	956.7	2,228.1	1,710.4	1.085
2004	August			1,856.1	1.000

Index = 0.6 Material + 0.4 Labor

Appendix E

Location Factors

Table E–1. Location factors

	1	2	3	4
Climate/Terrain	Benign	difficult	difficult	extreme
Gas Transmission	Present	present	no	no
Fresh Water	Present	present	no	no
Ship Loading	Present	present	no	no
Employee Housing	Present	present	no	no
Labor Costs	Low	high	high	high
Relative Capex	1.000	1.155	1.562	2.250
Relative Opex	1.000	1.139	1.520	2.039
Examples	U.S. Gulf	Urban Australia	Remote F.E.	Offshore
	Canada	New Zealand	Remote Australia	Arctic
		Developed F.E.		
		Middle East		

Note: Location factors were developed through U.S. DOE studies relative to U.S. Gulf. U.S. Department of Energy. "Assessment of Cost Benefits of Flexible and Alternative Fuel Use in the U.S. Transportation Sector," Technical Report Three, *Methanol Costs*, November 1989.

Appendix F

Methodology for Economic Analysis

The methodology described was devised by ICI PLC in order to evaluate all of the diverse routes to the production of ethylene from any feedstock. The methodology has been published by Stratton and others.[1] It is generally applicable for energy-intensive industries. The objective is to reduce the underlying economic variables to a fixed-variable relationship, with the variable being the feedstock of interest, in this case the price of natural gas.

The fixed part of the equation represents the return on capital, independent of tax considerations, together with all the nonfeedstock operating costs.

Capital

The capital costs are for greenfield projects completely isolated from other facilities. All the costs associated with utilities (unless otherwise accounted) are allowed for in the capital cost. Some plants require small amounts of power. This is considered as an import.

Capital is estimated using published information and using the location factors and Nelson-Farrar Indices. It is adjusted to U.S. Gulf Coast site costs and late 2004 costs for all processes.

Scaling has been done using the exponent method, namely:

$$\text{Capital of Plant [1]/Capital of Plant [2]} \\ = \{\text{Capacity of Plant [1]/Capacity of Plant [2]}\}^{n} \qquad (F.1)$$

where n is a constant with a typical value around 0.7.

Capital recovery factors

For a plant with a capital cost of C_0, the plant investment cost, C, capitalizes the return on investment during construction of the plant.

$$C = Co \sum_{S=0}^{p} a_s \,(1+i)^{p-s} \tag{F.2}$$

where

- a_s represents the breakdown of capital expended over the construction period,
- p is the first year of production,
- s is a general year of the project, starting at s = 0; construction is complete at s = p,
- i is the return on investment, and
- a is given in table F–1.

Table F–1. Values for a_s

Const. Period	Year 1	2	3	4	5
a_s					
a_0	100%	50%	30%	17%	4%
a_1		50%	45%	32%	14%
a_2			25%	26%	32%
a_3				25%	36%
a_4					14%

The general DCF equation can be written:

$$C = \sum_{r=1}^{N} \left((R_r - FC_r - VC_r)/(1+i)^r\right) \tag{F.3}$$

where

- r is the production year,
- N is the final production year,
- R_r is the total product revenue in year r,
- FC_r is the fixed costs in year r, and
- VC_r is the variable cost in year r.

This equation is simplified by assuming that there is no buildup to full production, and full production is achieved as soon as construction is complete. This is followed by N years of full production. Hence:

$$C\,(1 + i) = (R_r - FC_r - VC_r) \sum_{r=1}^{N} 1/(1 + i)^r \qquad (F.4)$$

This is rearranged to give:

$$R_r = FC_r + VC_r + K(1 + i)\,C \qquad (F.5)$$

where K is the sum of the geometric series:

$$K = i\,(1 + i)^N/[(1 + i)^N - 1] \qquad (F.6)$$

Values of K for various values of i and N are given in table F–2.

Table F–2. Values of K

N	10	15	20	25	30
interest (i)					
5.00%	0.1295	0.0963	0.0802	0.0710	0.0651
7.50%	0.1457	0.1133	0.0981	0.0897	0.0847
10.00%	0.1627	0.1315	0.1175	0.1102	0.1061
12.50%	0.1806	0.1508	0.1381	0.1319	0.1288

The capital recovery factor (K_o) is then:

$$K_o = K\,(1 + i)\,C \qquad (F.7)$$

From F.2, one gets:

$$K_o = K\,(1 + i)\,C_o \sum_{S=0}^{p} a_s\,(1+i)^{p-s} \qquad (F.8)$$

Values for the return on capital (ROC) or K_o/C_o are given in tables F–3 and F–4 for a royalty-free basis and one encompassing a 2% royalty to the process licensor, respectively. Table F–3 has been used for typical nonprocess items (pipelines, ships, etc.), and table F–4 for licensed processes.

Table F–3. Values for return on capital (royalty-free basis)

1-Year Construction					
N interest (i)	10	15	20	25	30
5.00%	13.60%	10.12%	8.43%	7.45%	6.83%
7.50%	15.66%	12.18%	10.54%	9.64%	9.10%
10.00%	17.90%	14.46%	12.92%	12.12%	11.67%
12.50%	20.32%	16.96%	15.54%	14.84%	14.49%
15.00%	22.91%	19.67%	18.37%	17.79%	17.51%
2-Year Construction					
N interest (i)	10	15	20	25	30
5.00%	13.94%	10.37%	8.64%	7.64%	7.00%
7.50%	16.25%	12.64%	10.94%	10.01%	9.44%
10.00%	18.80%	15.19%	13.57%	12.72%	12.25%
12.50%	21.59%	18.02%	16.51%	15.77%	15.39%
15.00%	24.63%	21.14%	19.75%	19.12%	18.83%
3-Year Construction					
N interest (i)	10	15	20	25	30
5.00%	14.61%	10.87%	9.05%	8.00%	7.34%
7.50%	17.74%	13.79%	11.94%	10.92%	10.31%
10.00%	20.99%	16.96%	15.15%	14.21%	13.68%
12.50%	24.66%	20.58%	18.85%	18.01%	17.58%
15.00%	28.77%	24.69%	23.06%	22.33%	21.99%
4-Year Construction					
N interest (i)	10	15	20	25	30
5.00%	14.88%	11.07%	9.22%	8.15%	7.47%
7.50%	17.39%	13.52%	11.71%	10.71%	10.11%
10.00%	20.58%	16.62%	14.85%	13.93%	13.41%
12.50%	24.17%	20.18%	18.48%	17.66%	17.23%
15.00%	28.20%	24.21%	22.61%	21.90%	21.56%
5-Year Construction					
N interest (i)	10	15	20	25	30
5.00%	14.71%	10.94%	9.11%	8.06%	7.39%
7.50%	17.61%	13.69%	11.85%	10.84%	10.23%
10.00%	20.91%	16.89%	15.09%	14.16%	13.63%
12.50%	24.66%	20.58%	18.85%	18.01%	17.58%
15.00%	28.87%	24.78%	23.15%	22.42%	22.07%

Table F–4. Values for return on capital with 2% royalty

1-Year Construction					
N	10	15	20	25	30
interest (i)					
5.00%	13.87%	10.32%	8.59%	7.60%	6.97%
7.50%	15.97%	12.42%	10.76%	9.84%	9.28%
10.00%	18.26%	14.75%	13.18%	12.36%	11.90%
12.50%	20.73%	17.30%	15.85%	15.14%	14.78%
15.00%	23.37%	20.06%	18.74%	18.15%	17.86%
2-Year Construction					
N	10	15	20	25	30
interest (i)					
5.00%	14.22%	10.58%	8.81%	7.79%	7.14%
7.50%	16.57%	12.89%	11.16%	10.21%	9.63%
10.00%	19.17%	15.49%	13.84%	12.98%	12.50%
12.50%	22.02%	18.38%	16.84%	16.09%	15.70%
15.00%	25.13%	21.56%	20.15%	19.51%	19.20%
3-Year Construction					
N	10	15	20	25	30
interest (i)					
5.00%	14.32%	10.65%	8.87%	7.85%	7.19%
7.50%	16.92%	13.16%	11.39%	10.42%	9.83%
10.00%	19.84%	16.02%	14.32%	13.43%	12.93%
12.50%	23.08%	19.27%	17.65%	16.86%	16.45%
15.00%	26.68%	22.90%	21.39%	20.71%	20.39%
4-Year Construction					
N	10	15	20	25	30
interest (i)					
5.00%	14.61%	10.87%	9.05%	8.00%	7.34%
7.50%	17.26%	13.42%	11.62%	10.63%	10.03%
10.00%	20.23%	16.34%	14.60%	13.70%	13.19%
12.50%	23.54%	19.65%	18.00%	17.20%	16.78%
15.00%	27.21%	23.36%	21.82%	21.13%	20.80%
5-Year Construction					
N	10	15	20	25	30
interest (i)					
5.00%	15.00%	11.16%	9.29%	8.22%	7.53%
7.50%	17.96%	13.96%	12.09%	11.06%	10.44%
10.00%	21.33%	17.23%	15.39%	14.44%	13.90%
12.50%	25.15%	20.99%	19.23%	18.37%	17.93%
15.00%	29.45%	25.28%	23.61%	22.87%	22.51%

Fixed operating costs

Working capital. Rather than capitalize the working capital and handle it with the project capital (Stratton), the working capital is treated as an annual operating cost. The reasoning behind this is that working capital is normally borrowed against the business and is fully recovered at the end of the project. The outgoings are the interest on the debt. The value of working capital can be taken as 5% of the plant capital or 30 days of stock. The latter is generally smaller than the former and was used when sufficient data permitted its calculation.

Labor, maintenance, and administrative costs. As a general rule, labor and maintenance were each charged at the rate of 3% of the capital per annum. For labor, this included both direct and indirect labor costs. For maintenance, this included both materials and labor. For large capital cost plants, such as LNG and GTL, lower values based on information of the manning levels at hand were used. Typically these were 1.5% of the capital as an annual charge. Over the past decade, many companies have made attempts to reduce the operating labor and maintenance charges. Labor can be reduced by extensive computer control. However, the success (or otherwise) in reducing the maintenance charge is difficult to quantify. Several operations have suffered major problems that were claimed to be due to the cutbacks in maintenance costs. Administrative costs are basically insurance and local land taxes. A value of 1.5% of the fixed capital as an annual charge was used.

Catalysts and chemicals. Most plants require some chemicals for water treatment purposes. Catalyst charges are based on a three- to five-year turnaround.

Variable operating costs

Gas usage. For most of the cases considered, the purchase of natural gas was the only variable of interest. This was estimated from published data, and allowances were made for the following:

- Most published data concern ISBL details. This does not include the use of gas for production of utilities (steam, power, water, etc.).
- Proprietary data tend to give an operational design gas usage for a new plant operating under ideal conditions. This was adjusted by allowing for normal operational problems and for catalyst deterioration.

Other variables. For some processes, other feedstocks are an important input (DRI). Also for some, relatively small amounts of imported power are required. In both cases, typical average values were used.

References

1. Stratton, A. "A Simplified Method of Calculating Product Cost," Technical Note 3, Economic Assessment Service, IEA Coal Research, London, 1982.

Index

A

D

E

F

G

H

I

J

K

L

M

N

O

Q

R

S

T

U

V

W

X

Y

Z